Maritime Radio and Satellite Communications Manual

Ian Waugh

WATERLINE

Published by Waterline Books
an imprint of Airlife Publishing Ltd
101 Longden Rd, Shrewsbury, England

ISBN 1 85310 471 X

A Sheerstrake production.

A CIP catalogue record of this book
is available from the British Library

The following Trade Marks (TM)/Service Marks (SM) are use in this book:
FleetNET   TM/SM of Inmarsat
SafetyNET   TM/SM of Inmarsat
IRIDIUM   TM/SM of Iridium, Inc.
B-Sat, C-Sat and *M-Sat* — service marks of BT Inmarsat, for INMARSAT-B, INMARSAT-C
and INMARSAT-M services respectively, through Goonhilly Coast Earth Station.

Printed in England by Livesey Limited, Shrewsbury

Acknowledgements

A large number of individuals in many organisations have provided me with help and information for this manual. People from the following organisations all played a part in helping me to find all the pieces which, hopefully, will make this first edition of the Manual one which you, the reader, will be pleased to have obtained.

Alden Electronics, Maryland USA;
American Radio Relay League;
Amsat-UK, London;
AT&T International Communication Services;
Department of Transport & Communications, Australia;
BBC World Service, London;
BT Inmarsat, London;
BT Coast Station Operations, London;
Department of Communications, CANADA;
The Canadian Coast Guard service;
The Canadian Radio Relay League;
Cimat SpA, Italy;
COMSAT Mobile Communications, Maryland, USA;
Cospas-Sarsat Secretariat, London;
Federal Communications Commission, USA;
Global Maritime Radiotelephone Service, London;
HWH Electronics, St Petersburg Beach, Florida;
ICOM UK;
INMARSAT, London;
International Maritime Organisation, London;
Department of Transport, Energy and Communications, Ireland;
Irish Radio Transmitters Society;
KFS World Communications, California;
MAT Equipment Antennas Marines, France;
McMurdo Ltd, Portsmouth, England;

Motorola Satellite Communications, Arizona;
New Zealand Association of Radio Transmitters;
Radio Frequency Service, New Zealand;
Portishead Radio, Somerset, England;
Radio Society of Great Britain;
Radiocommunication Agency, LONDON;
Radio Canada International;
Radio France International;
Royal Yachting Association, England;
Shakespeare Antennas, USA;
Department of Posts & Telecommunications, South Africa;
South African Radio League;
St. Katherine Yacht Haven, London;
STC International Marine, England;
Swedish Telecom Radio;
Swiss Radio International;
Telecom Mobile Radio Ltd, New Zealand;
Telecom NZ International Ltd, New Zealand;
Telstra Maritime Services, Australia;
Trimble Navigation, California & England;
TRW Space & Electronics Group, California;
UK Department of Transport (Marine Directorate), London;
V-Tronix Aerials, England;
Waterway Communications System Inc, Indiana;
Wireless Institute of Australia,
WLO Mobile Marine Radio, Alabama;
World Radio TV Handbook;
Wray Castle College, England;

I have not named people individually, but those who provided help in reviewing my draft scripts know themselves who they are and, I hope, realise how much their input was (and is) appreciated.

Two people I will name for the considerable effort they had to contribute: Margaret, who typed-up my notes, and then re-typed umpteen times as I revised text — and didn't once complain about the jobs which were not being done around the house; and Peter Coles, my editor, who took all the bits of the jigsaw and pulled them together into the single product — whilst coaxing me into maintaining progress to meet our deadline. And there's an illustrator out there somewhere who did wonders with my scribbled 'Figs ...'.

Contents

Introduction

The Maritime Radio and Satellite Communications Manual is designed to help mariners the world over to understand, and to get the best from, radio and satellite communications — wherever they may be.

It explains the way radio and satellites work and the way the radio frequency spectrum is divided up, internationally and nationally, amongst the multitude of service providers and users.

Because maritime radio and satellite systems are universal, equipment used to contact Aberdeen will also connect you with Zanzibar. You can use the same radio off Boston, Massachusetts and Boston, Lincolnshire.

Yachts and power boats, fishing vessels, ocean liners, oil tankers, bulkers, container ships and oil rigs all use the same radio and satellite facilities. They get their weather messages, make telephone calls and send telex messages over the same systems and ultimately, when necessary, make contact with search and rescue authorities anywhere in the world, using the same distress and safety procedures.

The procedures used are the same for radio officers, yachtsmen, fishermen, navigators and owner operators, and for radio medics on offshore installations and Coastguard, Navy, Air Force Rescue and Lifeboat men.

Although all mobile marine radio and satellite equipment works into the same shore-based services, some regulations regarding performance and durability of equipment differ for 'voluntary' fittings (like 'pleasure craft') and compulsory fittings. Operators qualifications and examinations also differ, country by country, and for voluntary and compulsory fittings. The need for an operator certificate also changes according to the type of equipment you use and whether you want to contact home stations, or foreign. Some radio channels are used differently from one country to another, so equipment which lets you use *most* channels around the world might not work on *all*.

This Manual covers inter-ship communications, port/harbour and ship movement services, radiotelephone and radio telex services, and maritime business (private) radio channels. It covers broadcast services, including distress and safety facilities, weather and navigation warnings, weatherfax and the broadcasts of 'Maritime Safety Information' under the Global Maritime Distress and Safety Service.

Marine VHF equipment, MF/HF Single Side-band and Ship Earth Stations for the Inmarsat system are all explained — as are the Mobile Satellite Systems of the future.

The Manual also covers alternatives to the normal global marine radio services. Cellular systems (maritime, and land mobile systems used by mariners) and the use of Amateur Radio in the 'Maritime Mobile' environment are also explained.

Regulatory aspects affecting services and equipment are explained and finally, the Manual explains how to choose the most appropriate equipment to meet your own particular communication requirements.

Boxed sections are used to elaborate some point of text or to explain another pertinent point, without interfering with the main flow of text.

Finally, we need the help of you the reader to shape this Manual to meet your requirements. As each new edition is published, we would like to know that it covered all your needs in the way that you would like to see it. A feed-back form has therefore been provided at the back of the Manual, so that you can record your opinion on the various chapters and send it back to the publishers. Your help in providing this feed-back would be gratefully appreciated.

Chapter 1

Radio Communications Explained

Introduction
This chapter briefly explains the need (or desire) for radio, and how radio transmitters and receivers work.

It covers the basic operation of transmitters and receivers; modulation and demodulation; filtering and amplification; radiation (from the transmitter antenna); and signal capture by the receiving system.

The different requirements for transmit and receive antenna are explained as is the role of the power supply unit.

Block schematic diagrams are used to support the text and as an aid to understanding the principles of radio transmission and reception.

Why Radio?
The twentieth century saw a real radio revolution. It opened with Guglielmo Marconi, the Italian physicist, extending earlier experiments in communicating without wires (wireless) to communicating across the Atlantic. The century closes with the full implementation of the Global Maritime Distress and Safety System (GMDSS), in 1999.

The fact that Marconi had a rather grand yacht might have inspired his research. He would have known the limitations of sound and visual communications. Megaphones, bells and whistles; flags, lights and flares all have there place, but they *don't* have *distance*.

Sound-waves, which we use for normal voice communication, rely on the atmosphere around us to carry them from mouth to ear. The vocal cords cause the air to vibrate and these vibrations travel outwards, gradually becoming weaker, until they cannot be heard any further.

Visual communication relies on light waves and on our own ability to recognise them. When it comes to reading the written message, we are all limited to *very* short distances. Yet light waves travel from galaxy to galaxy, apparently without hindrance!

Radio waves are like light waves. They *travel* at the speed of light and are *'electromagnetic'*. They possess the same infinite capacity for travel through free space, in any direction. Unlike sound waves, they *do* go the distance. So that is *Why Radio*.

Radio waves, on their own, are of limited value. We can neither see nor hear them, but what we can do is to arrange for our voice or visual message to hitch a ride, piggyback style, on one of those radio waves which we know can go *our* distance. The ride starts in the radio transmitter itself and ends with the radio receiver and enjoys a number of adventures on the way.

Radio Transmitters

Basic Operation

The *radio transmitter* is designed to produce a carrier-wave at radio frequency; to add your voice or other message to that carrier; and to release the combined *signal* into free space.

The transmitter itself is a series of electronic 'boxes', or stages (Fig 1.1), each designed to react to, or to manipulate, a supply of electricity. The result of that interaction is described below.

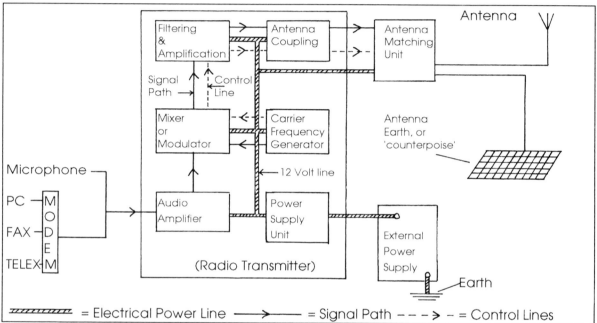

Fig 1.1 Radio transmitter arrangement – simplified schematic diagram

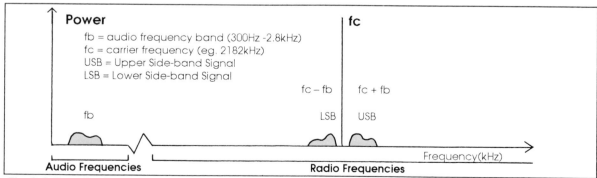

Fig 1.2 Result of combining audio frequencies with the carrier frequency in the modulator (mixer) stage

Power Supply Unit

The transmitter will normally receive its power from a generator or battery arrangement. Whatever that external source is, it is likely that an internal *Power Supply Unit (PSU)* will be provided as an integral part of the transmitting arrangement.

The PSU is designed to provide the various stages of the transmitter with a stable electricity supply, at the correct voltage for that stage. This stable electricity supply causes some stages to behave in a particular manner (as with the carrier frequency generator); whilst the supply itself is manipulated within other stages.

The electricity supplied to the various transmitter stages is not only there to 'make it work'; the *ambient voltages* and *currents* in various parts of the transmitter, the way these ambient levels behave and/or cause a particular circuit to behave, is crucial to the whole transmitting process.

Carrier Frequency Generator

The *carrier frequency generator* produces radio waves at the required radio frequency on which the voice or other message will be superimposed.

The basis of the carrier frequency generator is the oscillator circuit. The oscillator circuit is normally based on a *quartz crystal*, which is a particularly stable frequency source. The quartz crystal reacts to the voltage applied by the PSU by vibrating at a particular frequency.

The frequency generator circuitry in any particular transmitter is designed to cover all carrier frequencies in the band (or bands) at which the equipment is expected to be used (eg the Marine VHF band). The large number of carrier frequencies normally required in a transmitter is achieved by mixing the crystal oscillator frequency with the output from a *variable frequency oscillator (VFO)*, or by *frequency multiplication*, *frequency division*, or a combination of the three. The end result is the desired carrier frequency.

Audio Amplifier

The audio amplifier is the input stage for the message which is to be transmitted. It is the entry-point into your transmitter for the intelligence you want to pass on.

The message 'medium' can be your own voice, entering the transmitter via a microphone or it may be audio-frequency tones from a personal computer, fax machine or other device.

The microphone and audio input stage limits the band of audio frequencies to around 300Hz — 2.8kHz. The actual range of the human voice is much wider than this, but that small band is adequate for intelligible speech and understanding.

The audio amplifier provides a stable input level to the mixer, or modulator stage, where the message is combined with the radio frequency carrier-wave.

Hertz (Hz), Kilohertz (kHz) and Megahertz (MHz)

Audio and radio waves are cyclic and can be measured according to wavelength or frequency — where frequency is an expression of the number of complete cycles the radio wave goes through in one second of time, eg

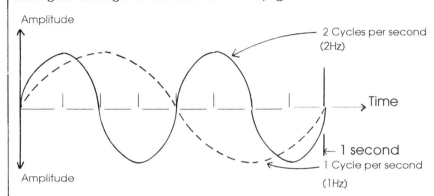

Older books will refer to the frequency as cycles per second, or c/s. The norm nowadays is to call those cycles 'Hertz', (abbreviated as Hz), after the German physicist of the same name.

Thousands of Hz are called kiloHertz (kHz); millions of Hz are called MegaHertz (MHz); which covers most maritime terrestrial radio frequencies.

Microwave radiation (and satellite communication frequencies) are measured in GigaHertz (GHz).

Voice frequencies range from about 300Hz — 6kHz, but speech can be readily understood with individual voice recognition, if the range 300Hz — 2.8kHz is captured. (The higher frequencies give our voice its individuality, rather than adding to clarity). So we ask our radio signal to carry only that portion of the voice signal which is really necessary — limiting the band-width required to less than 3kHz

Modulation

Mixing the voice (audio frequency) message with the (radio) carrier frequency in the modulator stage produces four distinct products (Fig 1.2), those being:

— The original band of audio frequencies (fb)
— the original carrier (radio) frequency (fc)
— the *sum* of the carrier frequency and the band of audio frequencies (fc+fb)
— the *difference* between the two (fc - fb)

The sum and difference combination are known respectively as the Upper Side-band (USB) and the Lower Side-band (LSB). As the USB and LSB are effectively mirror-images of each other — containing exactly the same information — only one side-band needs to be transmitted. This is the basis of Single Side-band (SSB) Operation.

The carrier frequency itself is purely a reference frequency and contains none of the message. If retained, however, it does use most of the available power and, for this reason, SSB equipment is designed to operate with the carrier 'suppressed'. A 'balanced modulator' effectively cancels out the carrier-wave, leaving only the two side-bands and the original band of audio frequencies.

Marine radio equipment is designed to automatically select the upper side-band for transmission, as this is the universally accepted mode on marine bands. Amateur radio equipment might offer the choice of USB or LSB modes of operations.

Filtering and Amplification

The remaining stages of the transmitter are designed to bring the desired signal — and only that signal — up to the output power allowed.

Filtering allows only the desired side-band (eg the USB) to get through. It also stops *harmonics* and other rogue signals, generated as a by-product of the complicated electronic circuitry of the earlier stages, from reaching the *antenna*.

Various stages of amplification, culminating in the 'final stage power amplifier' (PA stage), result in a strong signal being presented to the antenna. In marine radio, that signal is the upper side-band signal containing your message (that small band of voice frequencies, or the telex/fax/computer tones).

Antenna Coupling and Tuning.

The *antenna coupling* circuitry is tuned in sympathy with the Carrier Frequency Generator. The coupling circuitry should work like a sea toilet — allowing everything to pass smoothly through to the next stage with no 'flashback'.

The coupling circuitry is used to 'match' the transmitted frequency to the antenna on VHF sets, where the antenna length is constant. A matched frequency/antenna will allow the signal to pass easily into free space.

On MF/HF an additional unit — the *Antenna Tuning Unit* is needed to electronically alter the antenna length — fooling the coupling circuitry into believing that the antenna is of optimum length for the transmitted signal. As can happen with deficient sea toilets however, strong pressure from the outside can break down a return path and cause considerable trouble for any witness. If too great a mismatch occurs between the coupling circuitry and the antenna, you may well experience an electrical flashback — damaging your equipment possibly beyond repair.

Mismatch can occur when the transmitter is connected directly to the wrong length/type of antenna; where a MF/HF SSB transmitter is connected directly to a wire or whip antenna without an ATU between; where the wrong cable is used between the radio equipment and the antenna; or where intermediate connections become faulty.

Any mismatch will result in a portion of your signal being 'reflected' back into the equipment. If the level of reflected power equals or exceeds the 'forward' power then you are in trouble.

Voltage Standing Wave Ratio — VSWR

Reflected power is indicated by a Voltage Standing Wave Meter, which shows the ratio (*Voltage Standing Wave Ratio — VSWR*) of the power leaving the transmitter to that being reflected.

The ideal VSWR is a figure of one — or no reflected power (ie, all forward power being radiated). A VSWR of up to 1.5 would be considered efficient. Beyond 1.5 indicates a less than desirable arrangement which if it gets worse, could result in damage to your equipment.

The expected maximum VSWR from a fixed length antenna (eg a VHF antenna) should be quoted in manufacturers' literature. Check for a VSWR of 1.5 or better (ie, closer to Unity), when choosing a VHF antenna.

Similarly an ATU arrangement will specify the VSWR expected for MF/HF SSB. As actual antenna length will vary from one vessel to another — and the shorter the wire the more critical the set up — users are advised to have MF/HF SSB equipment professionally installed and tuned.

Although we call it 'reflected' power, the energy does not leave the transmitter and then return to do the damage. Any damage is caused by either a build-up of energy in the final stage which does *not* leave the set, or by the energy leaving too quickly (eg into a short-circuit).

Radiation

Having gone to all this trouble and expense, we need to release the signal into free space. This release is achieved by the antenna. The efficiency of the antenna arrangement — including ATU if used — will determine whether or not your message will have the power to get to the desired destination, or not.

The most efficient and practical antenna is a vertical wire, approximately half of one wavelength long, called a half-wave antenna. The transmitted signal passes to the antenna as alternating electrical current (alternating at radio frequency) and, where a good match exists, escapes into free space as *waves* (Fig 1.3a).

For the vertical wire and where no barriers exist close by, the radiation is omni-directional. This is particularly significant for sea going vessels, as any directional arrangement for ordinary radio transmission in a mobile environment can be severely limiting! (Although this is ideal for satellite systems).

With antenna length as a critical factor for efficient radiation, the limitations in a marine environment can easily be seen.

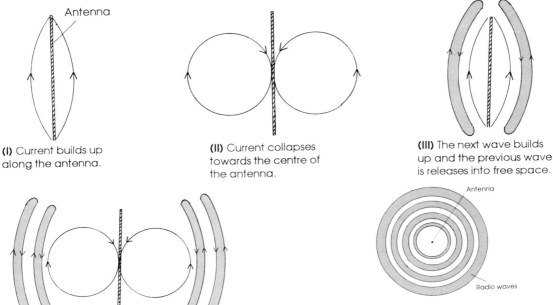

(I) Current builds up along the antenna.

(II) Current collapses towards the centre of the antenna.

(III) The next wave builds up and the previous wave is releases into free space.

(IV) As successive waves build up and collapse, they are radiated into free space at the radio frequency rate, ie – a 2182kHz signal releases waves at the rate of 2,182,000 per second.

Seagull's eye view of antenna and radio waves. The radio waves leaving the antenna resemble doughnuts, except that they grow bigger and wider as they move further away. The *total* power in each wave remains the same, but is shared more thinly as the wave becomes wider – resulting in very weak signals at the receiver.

Fig 1.3a Radiation pattern of a half-wave antenna.

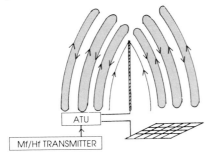

The quarter-wave vertical/whip antenna (with counterpoise), results in the same radiation pattern as the half-wave antenna – omni-directional and covering a wider/higher area the further it travels from the antenna.

Fig 1.3b Quarter-wave antenna

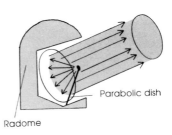

The microwave 'dish' used for most maritime satellite systems is a *directional* antenna and is pointed towards the satellite at all times. The radiation pattern resembles a torch beam – directional, but widening to some extent as it moves further away from the antenna.

Fig 1.3c Microwave radiation

VHF Antenna

In the maritime mobile VHF transmitter band of 156 — 162MHz, a half-wave antenna would be just under one metre (about three feet long, with a difference of only a few per cent from one end of the frequency spectrum to another). Antenna efficiency (represented by a VSWR of 1.5 or better) across the band is therefore relatively easy to achieve, with any matching required being done within the radio set itself.

MF/HF Antenna

MF/HF Antenna efficiency for a single side-band set operating in the 2MHz radiotelephone and radio telex bands, and in the HF bands of 4MHz — 25MHz, is a different proposition altogether.

A half-wave antenna at 25MHz (the highest Marine HF band) would be about 6m (20ft) long. Not too difficult to achieve, even on power-boats with no mast. At 2182kHz however, the ideal length would be 68m (over 200ft). Such elasticity is not common in wire antennas!

To make things even more awkward, the half-wave antenna should be sited about 5 wavelengths high, for maximum radiation efficiency. For the 25MHz (12 metre) signal, the mid-point of the antenna would ideally be 60 metres above sea-level. To raise the 2MHz RT band antenna (wavelength 400ft+) to the desired height would require a hot-air balloon, or a highly proficient flyer of kites! Fortunately, there are two steps which can be taken to make the MF/HF antenna more practical.

The first is to 'cheat' the system and reduce the actual length requirement to a quarter-wave. The second is to use an 'antenna tuning unit' (ATU — also, and more correctly known as an *Antenna Matching Unit* or *Antenna System Matching Unit*) between the transmitter output and the antenna wire proper.

The ATU is a compact unit which effectively becomes part of the wire antenna and should be capable of *electrically* adjusting the 'length' of the antenna, by the amount needed to match the various frequencies used with the transmitter. There is some loss of efficiency even with the best ATU, as some power which would otherwise have been radiated is lost within the ATU itself. In addition, the more an ATU has to lengthen an antenna, the more expensive it becomes. That is where one benefit of reducing the requirement to a quarter-wave antenna is felt.

The second benefit is that the antenna no longer has to be elevated into the skies in order to work properly. Fig 1.3b shows the quarter-wave arrangement, which operates as described below.

The antenna feed is split, with one end going to the vertical wire (or whip) and the other going to 'earth'. The earth then acts as a reflector, or counterpoise, to the radiating wire — effectively doubling its length. The vertical wire now only has to represent one-quarter wavelength, not a half wavelength. The antenna also radiates efficiently, even at near-sea level.

Water provides an excellent conductor to earth, giving mariners an advantage over land-based users. (As mariners are usually more restricted for space than those ashore, perhaps this is fair compensation?) An earth arrangement which makes maximum use of the sea is therefore used.

The down-side comes where MF/HF SSB is being retro-fitted to wooden hulled, glass-fibre or cement boats, where making a *suitable* earth connection can cause considerable upheaval. (See Chapter 5)

Radio Receivers

Receiver Basic Operation

The radio receiver has the job of recognising our own signal from within the multitude of other signals being picked up by the antenna at any one time. It then has to separate our message from the radio frequency signal which carried it, piggyback style, from the transmitter.

The arrangement shown on Fig 1.4a covers the basic operation of a radio receiver.

All waves passing through free space in the vicinity of the antenna will induce electrical currents into that antenna. The currents will be very small but will, nevertheless, affect the input stage of the receiver.

The first stage, the *Radio Frequency Amplifier*, increases the strength of selected signals and passes them into the *Mixer* stage. The *Local Oscillator* is tuned in sympathy with the RF Amp and passes an internally generated radio frequency signal to the Mixer. The RF Amp and Local Oscillator signals mix together and, as in the modulator stage of the transmitter, cause the sum frequency and difference frequency to be produced. One of those products is equal to the Intermediate Frequency (IF) of the receiver, and that IF is still modulated by the audio frequency message (fb). The *Intermediate Frequency Amplifier* again increases the strength of the signal, and passes it to the *Detector* stage.

The Detector separates the audio frequencies (fb — that small band carrying your message) from the IF carrier signal, and passes the audio band only to the next stage.

The *Audio Amplifier* increases the strength of that wanted audio message to a level suitable for the loudspeaker/handset or other device (eg fax/computer/telex). The following paragraphs explain the process in more detail.

The Receive Antenna

As with the transmitter antenna, the most efficient receive antenna would be of a length related to the wavelength (full/half) of the desired signal. Again, the antenna length could be electrically adjusted by an Antenna Tuning Unit.

With receivers however, the match is no where near as critical as with the transmit antenna, especially where a reasonably strong signal is being received. In fact, it is said, you can pick up a decent signal with a piece of wet string! A less-efficient receive antenna will make it difficult to pick up weaker signals however, but will not *damage* your receiver.

What happens is that *all* signals, those electromagnetic waves, from a multitude of transmitters, induce minute electrical currents into the antenna as they pass. This is because the antenna, as an electrical conductor, offers an easy path to earth for the electromagnetic waves. That path is through the Radio Frequency Amplifier, consisting of electrical circuitry which in its ambient state, operates at a fixed voltage level.

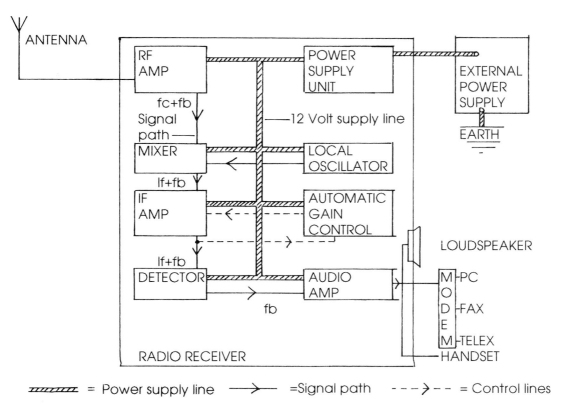

= Power supply line   ——→— =Signal path   - - -〉- - = Control lines

Radio receiver. We often think in terms of 'strong' and 'weak' signals being picked up by the receive antenna. In reality, unless you are very close to the transmitter, all received signals are weak. It is the receiver circuitry which transforms that very small signal – often measured in micro-volts – into one which seems 'strong'.

Fig 1.4a Radio receiver arrangement – simplified schematic diagram

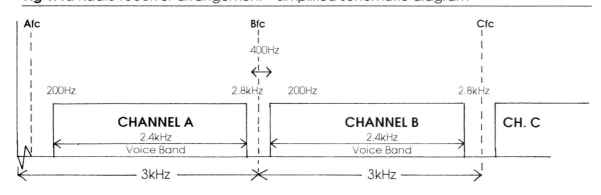

Afc = Carrier frequency for channel A   Bfc = Carrier frequency for channel B

Fig 1.4b Frequency separation between MF/HF voice channels

Radio Frequency Amplifier

The RF amp is tuneable across a range, or several ranges, of frequencies. The tuneability of the RF stage makes it sensitive to small electrical changes at a single frequency, or to a very small band of frequencies, at any one point of tune. All other signals in the antenna are thus reduced to a level at which they do not affect the receiver (they are 'tuned out').

In Marine VHF for example, the RF Amp is switched from one specific receive frequency to another as we change channels. The receiver reacts strongly to the chosen frequency and ignores all others.

The ability of the RF stage to pick up a very weak signal from the antenna is a measure of its *sensitivity*. The receiver must be sensitive enough to pass a signal containing the whole voice band for a given message — with reasonably constant amplification from the lowest to the highest frequency in that voice band — yet reject the adjacent signals (be *selective*).

The stages of the receiver following the RF Amp are a step-by-step conversion from radio frequency carrier, down to the audio frequency message.

Sensitivity and Selectivity

Modern receivers are designed to be ultra highly-sensitive — to be able to recognise the weakest of signals — and yet to reject other, unwanted signals close by the wanted signal. The ability to reject this *adjacent channel interference*, whilst remaining sensitive to the wanted signal, is known as *Selectivity*.

Adjacent channel interference should not be a problem in Marine VHF, satellite, or cellular radio services. These services operate on a fixed channel basis with standard frequency separation between each channel. It is therefore relatively easy for a manufacturer to design the equipment to reject adjacent channel signals without compromising sensitivity.

For communication receivers and other transceivers covering the more crowded MF/HF bands the problem is more acute. Channel separation in the MF/HF radiotelephone bands is very tight, typically leaving only 400Hz between signals of 2.6kHz bandwidth (Fig 1.4b); good selectivity is therefore much more difficult to achieve.

Mixer and Local Oscillator

The next step is to convert the carrier radio frequency (fc+fb) to Intermediate Frequency (If+fb). This is achieved in the Mixer stage.

The Mixer receives signals from the RF Amp, and from the Local Oscillator. The LO is tuned in sympathy, 'gang tuned', with the RF Amp and produces its own, unmodulated, radio frequency signal.

The signal produced by the LO is always different from the chosen RF signal by the amount of the Intermediate Frequency for that particular receiver.

As happened in the modulator stage of the transmitter, mixing the two (RF and LO) signals will result in both *sum* and *difference* frequencies being produced, one of which will be at IF and will be modulated by the audio band (fb).

eg
Modulated RF carrier *plus* LO signal = IF (+fb)
or
Modulated RF carrier *minus* LO signal = IF (+fb).

The design of the particular receiver will determine which arrangement is used. The output circuitry of the mixer and the subsequent IF stage are designed to operate at the intermediate frequency. Mixing the signals together in this way is called 'supersonic heterodyning' — an expression coined around 1920 and which, no doubt, sounded very futuristic at the time. Receivers using this process are called 'superhet' receivers and may be single, double or triple (etc) superhets, depending on the number of Mixer stages employed.

The choice of IF (the frequency itself and the number of Mixers/IF stages employed) determines the selectivity of the receiver in terms of both adjacent channel, and *image channel* rejection.

Image Channel Rejection

We have already seen that mixing the local oscillator signal with the received carrier frequency produces the intermediate frequency.

However, there is another frequency which, when mixed with the local oscillator frequency, will *also* produce the IF. This other frequency is above or below the local oscillator frequency by the same amount as the desired frequency eg

desired frequency	2182kHz
local oscillator	3982kHz (desired freq + IF)
intermediate frequency	1800kHz
image channel	5782Hz (LO freq + IF)
local oscillator	3982kHz
intermediate frequency	1800kHz

The receiver has to be sensitive enough to allow the desired signal through, but selective enough to reject that 'image' frequency. This selectivity is a function of the mixer stage (or stages) and the choice of Intermediate Frequency (usually different for each mixer stage).

Intermediate Frequency (IF) Amplifier

There is another advantage of the superhet principle and it is that, no matter what frequency the received signal is at, the intermediate frequency (or IFs) is constant in any one receiver. This makes the remaining amplification and detection processes easier. An IF Amp for example, only has to work at the same intermediate frequency (as modulated by that small band of audio frequencies) for all signals. The components which make up the IF Amp are therefore chosen to work best at the chosen intermediate frequency, without the compromise which would be necessary if a range of frequencies had to be handled.

This helps ensure consistent receiver performance, particularly with respect to amplifying the signal to a level suitable for the detector stage.

Automatic Gain Control

AGC is a feature of most modern communication receivers and is intended to compensate for varying levels of RF input to the IF stage. The AGC, when switched in, takes a sample from the output of the IF Amp (which will vary accord to the strength of signal received by the RF Amp) and will automatically adjust the amplification applied in the IF stage. If the level falls, the AGC will increase the IF amplification. If it rises, IF amplification will reduce. This ensures a reasonably constant input to the Detector stage.

Detector

The Detector stage separates the audio message from its intermediate frequency carrier, (ie. fb is separated from IF), allowing only the audio message (your voice or the computer/fax/telex tones) to pass to the Audio Amplifier.

Audio Amplifier

The Audio Amp is operator-adjustable. Users can therefore select the volume level required at the loudspeaker or handset, or adjust the level of input to the modem for PC, fax or telex messages.

Transceivers

The Transceiver is a radio set which combines both transmitter and receiver in a single unit. Transceivers are now the norm for Maritime Mobile VHF and most small-craft SSB sets.

Transceivers offer benefits in terms of ease of use, compactness, frequency stability and cost, over separate transmit and receive units.

Some circuitry (eg oscillators) are shared by both transmit and receive sections of the transceiver, making the combined unit more economical to produce and providing consistency of transmit and receive frequency stability.

Knobs and Whistles

Marine radio equipment, particularly that for small craft, is designed for ease of use. 'Knobs and whistles' are therefore kept to a minimum. This is not only because manufacturers do not expect yachtsmen, boat owners and fishermen to be technical experts, but because International Radio Regulations require certain functions to be automated, except for equipment which will be operated by fully trained technicians (eg Merchant Navy Radio and Electronics Officers).

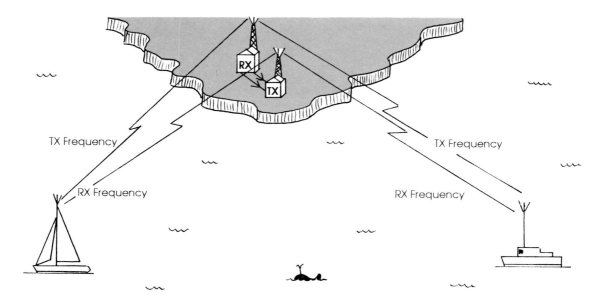

Radio Relay: The relay station ashore receives the signal on one frequency and automatically re-transmits on another. Two vessels which are outside of normal radio range of one another can therefore communicate with each other. (New Zealand uses this 'talk through' principle on designated marine VHF duplex channels.)

Fig 1.5a Radio relay

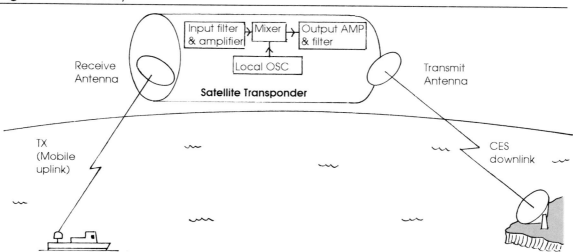

The satellite 'transponder' receives the signal from the vessel on one frequency. The received signal is amplified and mixed with the LO frequency, to produce the required frequency for transmission on the downlink to the Coast Earth Station. Another transponder does the same in reverse, for shore-to-ship signals.

Fig 1.5b Satellite relay

Amateur radio equipment is different. Many amateurs are more interested in experimenting than in exchanging messages, other than those of a technical nature. Amateur equipment will normally offer external, operator control of circuits which marine sets perform automatically. Even where the marine and amateur equipment is made by the same manufacturer, is the same size and *looks* the same, there will be different controls and displays.

Satellite and cellular radio equipment is even more automated than Marine VHF and SSB, and leaves little for the operator to do other than to switch on, select the station to work through, dial the required number and speak, (or press the data/fax send button).

Knobs and whistles specific to particular types of equipment are covered later in the appropriate chapter.

Control by Computer

Much modern equipment, like domestic appliances in every household, now has internal micro-processor control. Vast amounts of information on radio channels etc, can be programmed into the transceiver and recalled as required. Antenna tuning units 'memorise' the setting for particular frequencies. The satellite 'tells' the ship's Satcom unit to go to a particular channel, and it does. This is because the onboard Satcom equipment is programmed to respond to the satellite's instructions. Equipment can be switched on by a received signal in your absence and can record or print out your weather message or navigation warnings. It can ring bells to get you out of bed if it receives a distress message.

In addition to the internal micro-processor, personal computers are being used more and more — either to control the radio equipment or, more commonly, to 'look like' a telex machine or fax, or to send E-mail over the air and onwards to a shore subscriber. Configuration of equipment, including the use of personal computers, is covered in individual following chapters.

Summary

Radio (both terrestrial and satellite) communication, once the preserve of 'big ships' and requiring heavy, bulky, power-hungry equipment, is now available to all. Transceiver equipment is compact, easy to use and offers the choice of voice, fax, E-mail, data and radio telex — with low cost modems to interface with the ubiquitous PC.

Antenna tuning units make HF/MF SSB communication possible from the smallest of craft; and receivers are sophisticated enough to recognise the weakest of signals whilst rejecting the unwanted.

All we need is an understanding of the various types of equipment; a knowledge of what is available in terms of service for our own part of the world — and a better understanding of the way radio signals on different frequency bands travel from transmitter to receiver. That will ensure we get the best from radio and satellite communication; and that is the subject of the remaining chapters.

Chapter 2
Use of Radio Frequencies

Introduction

Chapter 1 explained how it is possible to generate and transmit a radio signal and how it is captured at the receive end. In this chapter we explore the different paths taken by radio waves of different frequencies when travelling from transmitter to receiver.

Different signalling methods, or *'classes of emission'*, are explained with the benefits of each. The relationship between power output and range for different frequencies is also examined.

The way the radio frequency spectrum is divided up amongst the international community, and between the multitude of service interests within that community, is also covered. We show how the maritime radio and satellite frequency allocations are interleaved with those of other radio interests.

Finally, those internationally agreed basic operating rules which enable us to communicate effectively with complete strangers, of various nationalities, wherever we are in the world, are demonstrated.

Electromagnetic Energy

Radio frequencies, as mentioned earlier, are a form of electromagnetic energy. The known electromagnetic energy spectrum, within which the radio frequency spectrum fits, is shown in Fig 2.1.

Radio waves — the lowest form of electromagnetic energy — have proved to be the easiest to manipulate. They are also the most likely to be affected by other forms of energy with which they make contact, planned or otherwise.

Fortunately, the boffins have worked out their natural behaviour; what influences their free passage through space and exactly how they are affected on the way. Because of this, we now know how to get the maximum benefit from the various forms of radio energy.

The radio waves within the radio frequency spectrum have been further sub-divided into various classifications, ranging from wavelengths of 1cm (30GHz) down to 100km (3kHz). The way these various classes behave — how they travel from one place to another and the range they can achieve over land — has been considered when deciding the best use to which they could be put. The passage of radio waves through free space is called *'propagation'*.

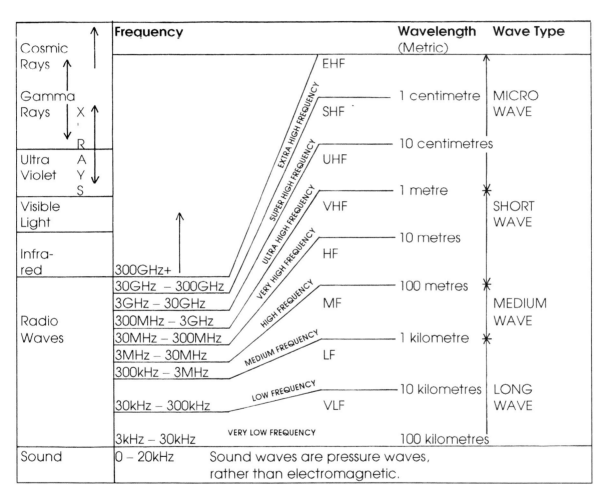

Fig 2.1 Electromagnetic energy spectrum and radio frequency spectrum.

Propagation of Radio Waves

There are three main methods of propagating radio waves, those being:

1 Direct or Space-wave — where radio waves travel virtually in a straight line and are unable to follow the curvature of the earth.

2 Ground-wave — where the radio wave clings to the surface of the earth even beyond the horizon.

3 Sky-wave — the most interesting method of travel, which uses the varying properties of the ionosphere (part of the upper atmosphere) to achieve long-range communications.

This is how they work.

Direct or Space-wave

Fig 1.3 showed how radio waves left the antenna and moved into free space. But the antenna is invariably close to the earth's surface and the signal, therefore, has more than 'free space' to contend with. The radiated signal quickly comes into contact with land or sea and, where frequencies at VHF and above are concerned, their passage is soon blocked. The range of a direct wave signal over the earth's surface is therefore very limited.

This method of propagation is considered to be 'line of sight' although the VHF *radio horizon* is actually slightly further than the visual horizon. The main factor in determining range in the Marine VHF band is the height of the transmit and receive antennas. If two yachts are communicating with each other and they both have their VHF antenna at a masthead height of 60ft, then they will achieve a range of around 20 miles. (Fig 2.2)

Either yacht will achieve greater range when communicating with a Coast Station, where the Coast Station antenna is on a headland or other high ground. Similarly, larger vessels with their higher masts and antenna mountings, achieve greater range than yachts and other smaller craft.

Increasing the power of a VHF transmission does not alter the range that can be achieved.

Satellite communications also use direct/space-wave, this time in the UHF band. In this instance the UHF signal passes directly through the atmosphere towards outer space with little hindrance. Even though the satellite is at a great height (around 22,200 miles/36,000 kilometres for Inmarsat satellites) the UHF signal can still go the distance. Fig 2.3(a) illustrates the propagation pattern for direct, or space-waves.

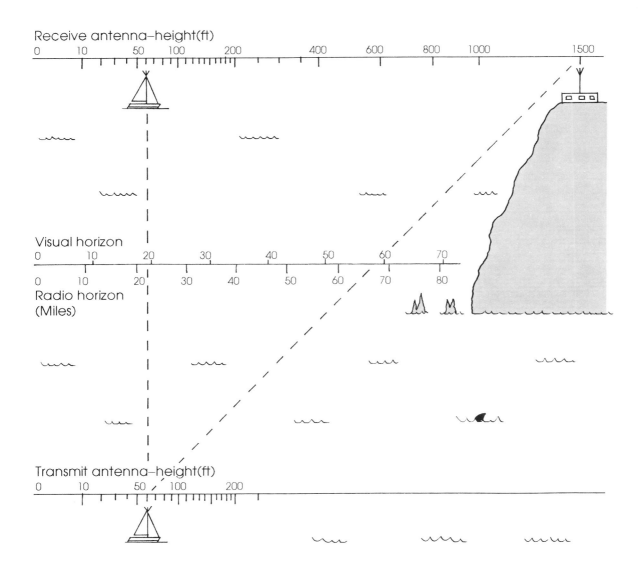

Fig 2.2 VHF radio horizon

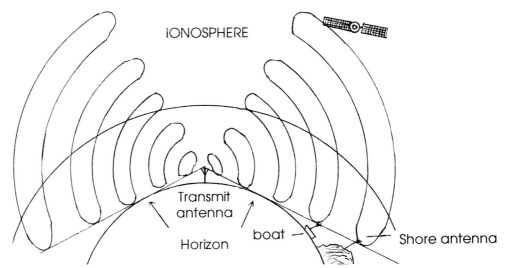

Direct waves can travel forever through free space, but range over the earth's surface is limited by the horizon. Direct/space waves are limited to the higher end of the frequency spectrum (VHF and above).

Fig 2.3a Direct or space wave propagation.
(Achieves 'line of sight' communication only.)

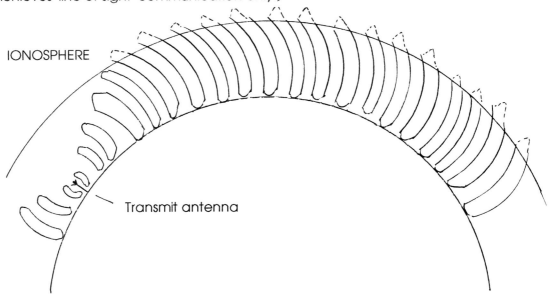

Ground waves follow the earth's curvature beyond the horizon. The range which can be achieved is related to the frequency and to the power output.

Fig 2.3b Ground waves

Ground-wave

Ground-wave propagation (Fig 2.3(b)) is the result of an interaction between the transmitted radio wave and the surface of the earth. Unlike space-waves, ground-waves actually follow the curvature of the earth and, depending on the wavelength and power output, can achieve world-wide coverage.

Ground-wave propagation is most important at the lower end of the radio frequency spectrum, from Medium-frequency down to Very Low-frequency. The longer waves of VLF and LF signals interact particularly well with sea and land, which effectively slow down the lower end of each wave whilst the rest continues through the atmosphere at its regular pace. The overall effect of this is that each wave bends towards the earth's surface as it progresses forward, so maintaining ground contact over great distances. (The top end of the wave, which continues skywards, is attenuated in the upper atmosphere and does not go off into space.)

At these low frequencies, the range that can be achieved by ground-wave propagation increases with power output. An efficient antenna arrangement is also necessary if the power from the transmitter is to be radiated, rather than lost in the system.

The power amplifiers and antenna arrangements needed for very long-range are large and preclude the use of VLF and LF transmission from most sea-going vessels. The use of these frequencies is therefore limited to the longer range *broadcast* services, including navigation systems such as Omega, Decca and Loran, which need to generate a consistent signal which will not be corrupted by atmospheric or other changes.

The Omega navigation system, for example, operates as low as 10kHz — the lowest practical radio frequency — and provides world-wide coverage with only eight transmitters. The Decca navigation system operates between 70 and 130kHz and Loran (Long-range Aid to Navigation) at around 100kHz.

The ability for signals to follow the curvature of the earth decreases as frequencies increase, so that at the Medium-frequency '2 Megs' maritime radiotelephone band, ground-wave range is limited to about 250 miles and cannot be extended significantly with more power. Frequencies in the maritime MF radiotelephone/telex band are expected to provide reasonably consistent coverage beyond VHF range, up to around 200 miles from the coast station concerned.

As we move into the High-frequency band, ground-wave coverage reduces even further, to only a few miles at the higher bands, becoming of limited practical use for maritime communications.

Sky-wave

It is in the maritime, broadcast and amateur High-frequency (short-wave) bands that sky-wave comes into its own. Unlike VHF, HF signals do not find an unimpeded path away from the earth's atmosphere and into space. This is because of the presence of 'ionospheric layers' in the upper atmosphere — this *ionosphere* having the ability to refract short-wave radio frequencies, bending them back towards the earth's surface.

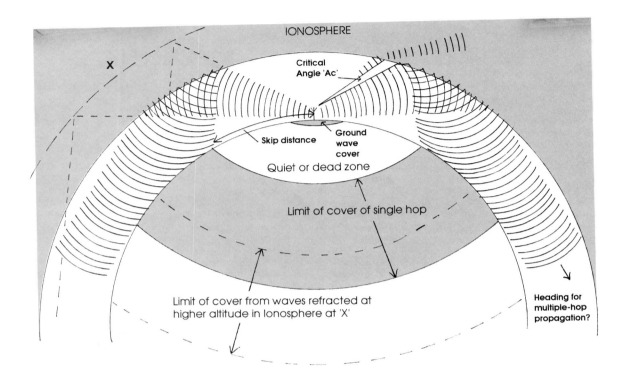

The radio wave is refracted by the Ionosphere and bends back towards the earth's surface. Ionisation varies according to the strength of the sun's rays, so the even return around the globe illustrated above is not possible in reality.

Critical Angle
There is an angle of entry into the Ionosphere 'Ac' beyond which the wave will not be bent enough to return to Earth. The sharpest angle which will return is known as the critical angle. Radio waves entering the Ionosphere more acutely will pass on through into outer space, albeit in a different direction to that of entry.

Refraction from higher levels
A radio wave that bounces off a higher layer of the Ionosphere as indicated at 'X' above, will return to cover a different part of the Earth's surface. Higher layers give longer range and a greater quiet zone.

Multiple-hop
Those radio waves that are by-passing the Earth's surface will meet the Ionosphere again. They may be refracted at the next point of contact with the Ionosphere, resulting in multiple-hop propagation and so achieving a greatly extended range.

Fig 2.4 Idealised sky wave propagation

The ionosphere gets the strength to do this from the sun, whose infra-red rays 'charge up' the ionosphere, endowing it with electromagnetic powers which affect the free passage of the shorter wave radio signals, particularly in the HF bands.

In the idealised representation shown in Fig 2.4, the omni-directional short-wave signal moves skywards until it reaches the ionosphere. The ionosphere refracts the signal, causing a portion of it to bend back towards the earth, which it meets some distance away from the transmitter.

Because the transmitted signal enters the ionosphere over a wide area, the returned signal also covers a reasonably wide area of the earth's surface.

The distance between the antenna and the first point of return is known as the *skip distance*. The area between the outer limit of the ground-wave and that first point of return is known as the 'quiet' or 'dead' zone. The signal has leap-frogged that part of the earth's surface and cannot be received in the quiet zone, even with the most sophisticated receive system. The area between the first point of return (skip) and the furthest returned wave represents the coverage attainable.

Sky-wave propagation would be very boring if that was all there was to it. It would also be very limiting. It would be like having a fishing rod which could not be cast less than 50ft and no more than 60ft; or a pair of binoculars which could only be focused at 200 — 300ft. Who would thank you for either?

Like a fisherman who chooses a different rod and bait, we can aim our radio to achieve different distances — or to pick up different prey — by selecting different frequencies within that HF band. The frequency we choose depends on the distance to be covered and the energy state of the ionosphere at the time communication is required.

Ionospheric Conditions

The sun is responsible for energising the ionosphere. The energy state of the ionosphere and the effect on HF propagation will therefore change according to the strength of the sun, and in particular with the level of infra-red rays passing through the upper atmosphere. The strength of those rays will change because of:

    1 Time of day or night.
    2 Geographical latitude.
    3 Season of the year.
    4 The varying level of radiation from the sun caused, for example,
    by sunspot activity.

Ionospheric Layers

The sun's infra-red rays cause electrons to fly free from the atoms in the upper atmosphere — leaving positively charged 'ions' — hence the name 'ionosphere'.

The varying levels of radiation passing through the ionosphere, because of the simple daily rotation of the earth, results in particular (and predictable) variations in the ionosphere. The most significant daily pattern is the energising of different ionospheric layers at various heights above the earth's surface. The ionospheric layer structure is illustrated in Fig 2.5.

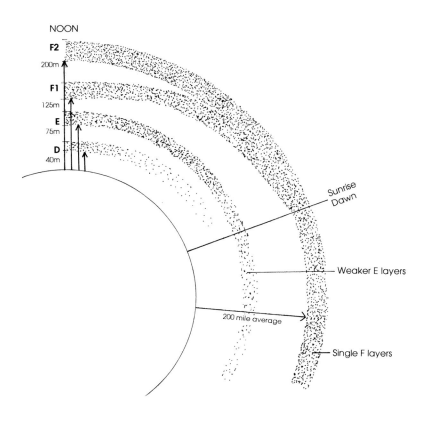

There are four layers of the Ionosphere, active at different times of the day and night and at different heights.

D = lowest = Most active around noon. Lowest layers therefore shortest skip and least range.

E = next lowest = Active during the day and to a lesser degree, at night.

F1 = second highest = Active throughout the day refracting higher HF bands.

F2 = highest = Active throughout the day, refracting highest HF bands.

F1+F2 combine at night to form a single layer, refracting the lower HF bands (as F layer is weaker during the hours of darkness, higher HF bands will pass through in the same way that VHF does during the day).

D layer when active absorbs lower (4/6MHz) signals. They cannot reach the E/F layers to be refracted, as happens at night.

Fig 2.5 Ionospheric layers (heights approximate)

Because the atmosphere gets thinner as we go higher, the free electrons at the outer edges of the ionosphere have difficulty finding another ion to neutralise. Ionisation is inclined to persist longer as we move further from the earth's surface.

Ionisation increases with the strength of the sun so that as we move towards mid-day ionisation approaches its maximum.

F and E-Layers

At the maximum point, two distinct F-layers are present, with F2 being anything up to twice the height of the F1, or the single night-time F-layer.

The E-layer is also most effective as a reflector during the day, as its charge increases.

D-layer

The lowest ionospheric layer is the D-layer, which is active from shortly after sunrise, to around sunset. Being at a much lower level however, the atmosphere is much more dense than the upper layers and free electrons have little difficulty re-combining with an ion and thereby neutralising it.

The effect of this is that, except for a short period around mid-day when ionisation is at its peak, the D-layer does *not* reflect signals — and actually absorbs those of longer wavelengths towards the lower end of the HF band and all frequencies below HF.

Geographical and Seasonal Variations

As the reflecting power of the ionosphere varies with the strength of the sun's rays, and the strength of the sun's rays varies according to latitude and season of the year, these two factors also have an effect on ionisation levels. The level of ionospheric charge reduces as latitude increases, and is lower in winter than in summer.

Perversely, the lower levels of winter charge reduce the negative effects of the D-layer, whilst other atmospheric changes associated with hot summer weather tend to cause more disruption. Sky-wave communication can often therefore be more consistent and rewarding in winter, when the need to communicate in an emergency might be more important than in summer.

Wavelength and Range

We saw earlier that low frequencies use ground-wave only, and that their range was dependent on power output. We also saw that very high frequencies (and above) achieve line of sight range only over land, but carry on through the ionosphere, unaffected, and off into space. Both therefore have reasonably consistent and predictable range.

The medium and high-frequency bands fall between those two extremes and need some thought when deciding which to use.

The question can easily be illustrated by considering the maritime MF and HF radiotelephone bands. The bands in question are at 2MHz, 4MHz, 6MHz, 8MHz, 12MHz, 16MHz, 22MHz and 25MHz — quite a selection — and their ability to propagate using sky-wave varies with the changes in the ionosphere.

Sunspots and Solar Storms

The level of radiation actually leaving the sun can also vary, in particular because of sunspots and solar storms. Sunspot activity follows an (approximate) eleven year cycle and, at its peak, considerably increases the options open to us for sky-wave propagation at higher frequencies, whilst reducing the options at lower frequencies.

The sunspot cycle is tracked regularly and the effects are predicted for us well in advance — so we know that 1994 is a 'low' and that conditions will now improve for a number of years.

Solar storms are different. As the storms flare up on the sun, they send out great bursts of energy which take only eight minutes to reach the earth. The result is a 'Sudden Ionospheric Disturbance' (SID) which initially disrupts communications at the higher frequencies.

If you are in HF contact and the signal quickly fades out, leaving only 'mush' across your HF band, that is probably a result of a SID. To regain contact you should try a lower frequency band.

For all practical purposes, the main ionospheric change we need to consider is the daily change — the effect of which is illustrated in Fig 2.6.

During the hours of darkness, when the D-layer has de-energised and so cannot stop the lower frequencies from reaching the E and F-layers, those lower frequencies get through and are refracted back to earth.

The higher frequencies are also passing through to the upper layers but, as those layers are not fully energised, the higher frequencies tend to pass right through, in the same way that VHF signals do during the day.

During daytime, the frequencies which will bounce back increase with the level of ionisation so that in the early afternoon the highest frequencies will be refracted.

It can be seen that there are always four or more frequency bands which will bounce back to earth. Those bands will be bent back at different heights and at different angles. Generally speaking, the lower frequencies will be bent back first and will therefore achieve shorter ranges overall than a higher frequency being used at the same time.

The figure used is a rough guide only, to demonstrate the pattern of frequency availability. The easiest way to learn which frequencies are best for any particular area is to listen regularly to a variety of coast stations, on different frequencies, and get a 'feel' for propagation.

Optimum Working Frequency Charts

A number of coast stations operating in the HF bands provide *Optimum Working Frequency (OWF)* or *Optimum Transmission Frequency (OTF)* guides. These guides predict likely best frequencies for HF contact over the coming months, for various sea-areas. The guides can be mailed or sent by radio teleprinter if you are already at sea.

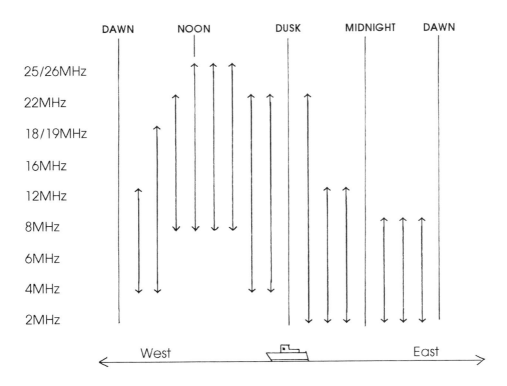

When considering which frequency to use, you should consider the time of day mid-way between yourself and the station you wish to call. On the North–South path the time will be the same as your own.

Communicating to the East or West will involve using the Ionosphere at a different state of charge — either ahead or behind that directly above your own position.

Regardless of time of day or night, the highest open frequency will cover the greatest distance and the lowest the shortest. Each band does cover a considerable range however, and there is much overlap.

1994 is a low point in the (approximate) 11 year sunspot cycle, so propagation will be limited at the higher HF frequencies. As sunspot activity increases during the latter half of this decade the higher frequencies will again perform well, both in range and time active.

Fig 2.6 Sky wave propagation – rough frequency guide

Fading: Change Up or Down?

It is useful to be able to recognise when a particular frequency is about to become unusable and whether to move up or down the bands when that happens. Fading is a good guide.

If your received signal appears to be steadily weakening, fading gradually into the mush, it is being absorbed by a strengthening D-layer. It is time to move up a band or two.

If you start to experience *irregular* fading — signal coming and going — it is probably because the signal is beginning to escape through the ionosphere as the E/F layer charge decreases. In that case, you need to move down the bands.

Propagation Headaches and Cures

When I was at Nautical College, training to be a Radio Officer, trying to learn about sky-wave propagation gave us all a headache. If this is your first time trying to learn about it, you too should have a headache by now. But there's no need to 'learn' it — just to be aware of *why* things change and *how*.

The best way to understand sky-wave propagation — once you know roughly how it works — is to get a short-wave receiver installed and start listening into the various stations across the HF spectrum. But it needn't be just a game. There is much valuable information flying through the air which a reasonably affordable set-up will give you access to without installing a transmitter as well, nor taking any special examinations.

Things like weather broadcasts (voice, telex or weather fax) and time signals (do you still use them for navigation?). You could monitor the amateur Maritime Nets and listen to the maritime amateur radio 'club'. You can listen into traffic lists and calls from various coast stations and consider which you would prefer to use, should you decide to fit a transmitter later on.

The range of services and providers using HF sky-wave propagation is covered in Chapters 5 and 6.

Frequency Allocation, Allotment and Assignment

The entire range of radio frequencies is not available to all users for any purpose they may choose, wherever they are on the globe or in space. Instead, the International Telecommunication Union (ITU) oversees a collection of conferences which decide how each radio frequency will be used, for what and by whom. The process involves the *allocation* of frequencies to services; the *allotment* of frequencies to areas or countries; and the *assignment* of frequencies to particular users by the allotted country.

The first part of the process is the allocation of frequencies to services, including, for example, the Maritime Mobile Radio Service, the Maritime Mobile Satellite Service and the Amateur Service.

The assignment process goes on to designate frequencies to services *within* the main categories of service including, for example, the Port Operations Service, Ship Movements Services and Maritime Radio Navigation Service.

Conferences also decide on *how* the frequencies should be used by the service to which they have been allocated. This includes the pairing up of frequencies to form a dual frequency channel, where one frequency will always be 'assignable' to a coast station and the other to ship stations.

Frequencies will be further specified as for use in, for example, the radiotelephone service, or for *narrow-band direct printing* (NBDP) — including Telex Over Radio (TOR) services or for some other specific use (eg NAVTEX).

The *mode*, or *class of emission* which can be used on a particular frequency is also specified. Some of the more common classes are listed below.

Class of Emission

The class of emission is shown as a 3-character group. The first character shows the *type* of modulation (eg. double side-band; single side-band suppressed carrier). The second describes the *nature* of modulation (eg. continuous wave — CW (Morse); analogue (voice); analogue (fax/computer tones). The third symbol describes the *form* the information to be transmitted will take (eg. telegraphy; telephony; television; facsimile). Some typical classifications used in the maritime world are:

Designator	*Which Means*	*Typical Use*
A3E	A = Double Side-band(AM) 3 = Analogue E = Telephony	Normal mode for 2182kHz Distress Calls
J3E	J = Single Side-band(AM) (suppressed carrier) 3 = Analogue E = Telephony	Normal mode for maritime MF/HF radiotelephone calls
F1B	F = Frequency Modulation 1 = Frequency Shift Keying (FSK) B = Telegraphy	Narrow Band Direct printing (NBNP) - NAVTEX; Digital Selective Calling (DSC)

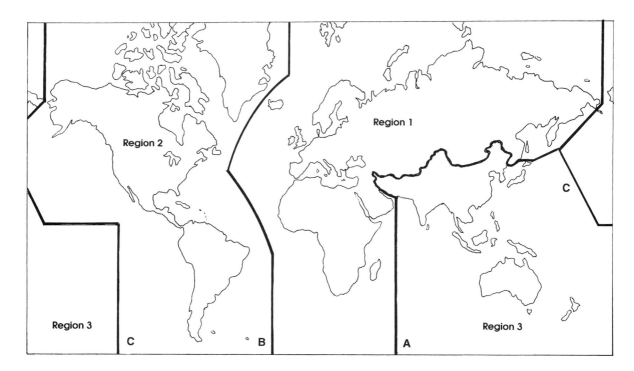

Fig 2.7 World Radio Regions

Regions and Areas

The world has been divided into three regions for the purposes of allocating radio frequencies. The map (Fig 2.7) shows the three world regions. These regions are used for allocating frequencies for all services, not just the maritime mobile or amateur radio service.

Region one includes Europe, Scandinavia (including Iceland), Africa and parts of the Middle-East (but not Iran) and Turkey — and most countries in the Commonwealth of Independent States (CIS). Region one also includes the Mongolian Peoples Republic and all sea and land mass north of The Russian Federation to the North Pole.

Region two covers the Americas, Greenland and all sea and land mass between line B in the East and C in the West, to the North Pole above Canada and, similarly, to the South Pole below South America.

Region three includes Iran, Afghanistan, Pakistan, India, South-East Asia, China, Australia and New Zealand and many of the Pacific Islands — being the area between line A in the West and line C in the East and on south to the Pole.

This regional structure may appear to have little significance in the maritime mobile radio and satellite bands, as seafarers are generally free to work any coast station they wish and can contact from wherever they are at sea.

The main significance for mariners of the regional structure, is where amateur radio is concerned. Certain amateur frequencies are only allowed to be used in particular regions and so when you are 'Maritime Mobile' you should observe local rules.

Although frequencies are allocated to particular services and, where the maritime mobile radio service is concerned, the allocation applies world-wide; the assignment of frequencies to individual coast stations is usually down to a single administration.

This allowed the US Federal Communications Commission to exercise its humour (could it really be coincidence?) when it allocated maritime VHF Channel 25 — a public correspondence radiotelephone channel — to the local Marine Operator services at *Madonnaville*, Illinois; *Convent*, Louisiana; and *Virgin Islands Radio!*

The short range nature of VHF means that Channel 25, and the other VHF channels, can be allocated to a number of stations in any one country and also to stations in many other countries with little fear of interference.

As the frequencies to be used for this channel are the same all around the world, we can then use the same piece of radio equipment to contact coast stations wherever we go.

Frequencies in other bands, including the MF and HF bands, are also shared by a number of coast stations in different countries. The different propagation factors in these bands, however, means that some interference is inevitable. Although this can be a nuisance at times, it also opens up the opportunity to contact different coast stations wherever you are at sea.

Simplex, Duplex and Semi-Duplex

Different types of service have different needs and those needs influence the way frequencies are allocated.

Port operations, ship movement services and inter-ship requirements are usually met with a single frequency channel. Both stations in a conversation will then transmit and receive on the same frequency and anyone else can monitor exactly what is being said by both parties. This is essential where safety of navigation at close-quarters is concerned.

Using a single-frequency radio channel means that only one party in a conversation can speak at a time — this being known as 'simplex' communication. (See Fig 2.8)

The VHF, MF and HF Public Correspondence channels — those on which we make our ship-to-shore radiotelephone calls, are two-frequency channels: one frequency for the Coast Radio Station and the other for the ship station.

Coast Radio Stations normally use separate transmit and receive installations, with the transmitter and receiver each having their own antenna. The CRS is therefore capable of 'full-duplex' working. Ships stations can also be fitted with full-duplex equipment, allowing them to talk and listen at the same time. An exchange between ship and shore is only full-duplex where both ends have this duplex facility. Full-duplex may not seem important for normal radiotelephone calls, but is essential for most types of data/facsimile exchanges.

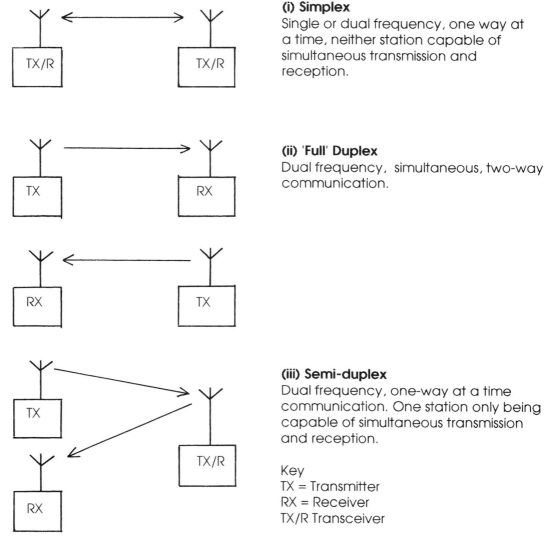

(i) Simplex
Single or dual frequency, one way at a time, neither station capable of simultaneous transmission and reception.

(ii) 'Full' Duplex
Dual frequency, simultaneous, two-way communication.

(iii) Semi-duplex
Dual frequency, one-way at a time communication. One station only being capable of simultaneous transmission and reception.

Key
TX = Transmitter
RX = Receiver
TX/R Transceiver

Fig 2.8 Simplex, Duplex and Semi-duplex communication

The most common arrangement for small craft on VHF is for the vessel to be fitted with a transceiver with a single antenna, which is capable of simplex operation only. Where this is the case and the shore end has a full-duplex arrangement, the configuration is known as semi-duplex.

Maritime Satellite communications are normally full-duplex — all telephone and telex channels being dual frequency and shore, space station and ship equipment being capable of simultaneous transmission and reception.

Basic Operating Rules

A number of basic operating rules and procedures have been established internationally so that we can get the best out of the radio frequency allocations available to us.

The two most significant items for mariners are:

1 The need for every transmission in the maritime, amateur and broadcast bands to carry an *identification signal*.

2 Standard procedures for distress, urgency, safety messages and for calling and working Coast Stations and other vessels.

Identification Signals

Anyone who is granted an amateur radio licence is allocated a personal *call-sign*. That call-sign must be used to identify all radio transmissions made by the individual transmitting on the amateur bands, including amateur-maritime nets. (Chapter 6).

Maritime licences differ from amateur licences in that the licence, the associated call-sign and other identifiers are normally allocated to the vessel on which the radio/satellite equipment is installed, rather than to the individual.

Transmissions from sea-going vessels on maritime mobile radio and satellite bands must also be identified but the method of identification used can vary according to the service being used. With the exception of Morse transmissions, which invariably use the allocated call-sign for identification, most transmissions in the maritime mobile radiotelephone bands use the vessel's name and/or call-sign for identification, eg.

'.... this is *Atlantic Drifter* call-sign KYCX ....'

Using the name *and* the call-sign identifies not only the vessel, but the nationality (see Appendix B for international call-sign allocations).

The name of the company operating a particular vessel can be added to the vessels name where it is needed to distinguish one vessel from another with the same name, eg

'... this is Universal Towing *Tugboat Annie* ....'

(Methods of identification used in satellite and other automated services are covered in the appropriate chapter.)

Distress, Urgency and Safety

The ability to gain access to important safety information and to communicate in an emergency, continues to be a key reason for fitting marine radio equipment onboard any vessel.

In order for emergency (Distress and Urgency) transmissions to be heard and understood, particular frequencies have been set aside, world-wide, and standard procedures have been agreed.

Distress, Urgency and Safety transmissions cover the following circumstances.

A **Distress** situation is where a ship, aircraft or 'other vehicle' is in *grave and imminent danger and immediate assistance is required.*
* 'Grave danger' means terminal — for the vessel and person(s) concerned —
* and the reason that 'immediate assistance is required' is because the danger is certain to result in the loss of the vessel and of life if rescue is not quickly forthcoming.

A yacht in mid-Atlantic, in good weather, with a fire below and out of control, would be in grave and imminent danger. The yacht is likely to sink and the people taking to the life-raft would undoubtedly require rescue as soon as possible. This would constitute a distress situation.

An **Urgency** situation is where there is *serious concern* over the safety of a ship, aircraft or person, but where any danger is not both grave *and* imminent. The type of assistance required may or may not require *rescue*, but could be for example take the form of *medical advice* or *medial assistance.*

A requirement for a tow into harbour to avoid getting into grave danger would also constitute an urgency situation.

Urgency situations, if not helped, can deteriorate into a distress case. Distress and urgency situations usually stem directly from problems onboard or with persons lost overboard. The exception to this would be an act of piracy against one vessel by another, which would also constitute a distress situation.

Safety transmissions are those concerning important navigational or important meteorological warnings. Safety transmissions are therefore concerned with warning *others* of potential danger, rather than communicating your own problems. Gale warnings, strong-wind warnings and dangers to navigation — including any malfunction to radio navigation systems, constitute safety messages.

Distress, urgency and safety transmissions consist of a signal, a call and a message.

The radiotelephone distress signal is the word 'Mayday', taken from the French *'M'aidez'*, meaning 'Help Me'. The Morse distress signal is SOS, which is also used on radio telex channels.

The radiotelephone urgency signal is 'Pan Pan', each word pronounced as the French *'panne'* which, appropriately, means 'breakdown'. The Morse/radio telex equivalent signal is 'XXX'.

The radiotelephone safety signal is the French word 'Sécurité', pronounced correctly in French, as 'SAY — CURE — E — TAY'. Used when messages concerning your safety, or security, are to be passed.

Distress, urgency and safety signals are spoken/keyed three times, followed by the appropriate call and message.

Distress Transmission

The following is an example of a complete radiotelephone distress transmission:

Distress Call:	'Mayday, Mayday, Mayday. This is *Albert Ross*, *Albert Ross*, *Albert Ross* call-sign Golf Uniform Lima Lima'
Distress Message:	'Mayday *Albert Ross* Position 28 degrees 12 decimal 3 North 32 degrees 25 decimal 5 West. Fire below, out of control. Require immediate assistance. Three persons onboard. Taking to life-raft, orange inflatable, no markings. Activating 406MHz EPIRB and Search and Rescue Transponder. Taking hand portable VHF onto life-raft, will maintain watch on Channel 16, on the hour, for five minutes over.'

Using this standard procedure on the correct frequencies will give a casualty the best possible chance of rescue. That is because the procedure is known all around the world and, even where English is not the first language, other stations know what to expect when they hear that first distress signal:

Station in Distress *(Casualty)*	*All Other Stations*
'Mayday, Mayday, Mayday'	All other stations must cease transmitting as a distress call and message are about to follow.
'this is *Albert Ross*, *Albert Ross*, *Albert Ross*, call-sign G-U-L-L'	*Albert Ross* is the name of the vessel. The call-sign distinguishes it from other *Albert Ross*'s and also shows the nationality
'Mayday, *Albert Ross*'	The distress message is starting, the next thing you will hear is the position.

'Position 28 Degrees 12 decimal 3 North 32 Degrees 25 decimal 5 West.'	This position is in latitude and longitude. It might have been a position relative to a known point of land. It could have been given as Decca, Loran or Omega co-ordinates. It might have been lat and long at, say, noon — followed by a course and speed since that time. It depends on the situation and the time available to compute a position.
'Require immediate assistance.'	For a Mayday call, the assistance asked is normally 'immediate'. Only in a lesser emergency would anything different (eg tow; medical advice/assistance) be likely to be appropriate.
'Over.'	You are invited to reply if in a position to help.

Can You Help?

If you are near to the distress position and able to offer assistance, you should reply immediately, taking care not to interfere with further transmissions from the vessel in distress, or other people ready to assist.

Your reply should take the following form:

'Mayday	The distress signal, once.
Albert Ross	The name of the vessel being answered, once.
this is *Gueniveve,*	Your own vessel's name
Gueniveve, Gueniveve	three times.
twelve miles South of	Your position and a
your position and	statement showing that
proceeding. ETA two hours.	you are going to help.
Will call on VHF	Your estimated time of
Channel 16 over.'	arrival at the scene of distress. Any other info *you* think will help.

If you intercept a distress call and are in no position to provide assistance yourself, you should not add to the radio traffic just to say you cannot help.

Mayday Relay

You might find yourself in a position where you have picked-up a distress message and are in no position to help (or can provide only limited assistance), but nobody else has responded to the vessel in distress. You can then get additional assistance by using the 'Mayday Relay' procedure on the original distress frequency, or on another frequency if appropriate.

An example of this could be a yacht in distress about sixty miles offshore, in thinly populated waters and out of range of the nearest Coast Station. Their distress message is broadcast on VHF Channel 16, and you are apparently the only one who hears it. Your own position is 35 miles offshore, between the distressed yacht and a Coast Station.

It will take you some time to get to the distress position — and that might be too late to help. If you are only carrying VHF radio, you will also be moving out of range of the Coast Station and possibly, any other vessels.

You need to re-broadcast the distress message on behalf of the casualty to ensure the best chance of rescue. This is the procedure.

Relay Vessel	*Other Listeners*
'Mayday Relay, Mayday Relay, Mayday Relay'	Someone is about to re-broadcast a distress message on behalf of a casualty.
'This is Yacht *Gueniveve, Gueniveve, Gueniveve.*'	Name/identification of the relaying station.
'Mayday Relay'	The message is about to begin.
'Following received from *Albert Ross*, call-sign Golf Uniform Lima Lima at 1345 UTC on VHF Channel 16, Begins.'	State who you received the message from, at what time, and on which frequency/channel.
'Mayday *Albert Ross* In position (etc)'	The complete distress message (without the original call) is repeated, without further editing/abridgement.
'Over'	An invitation to other listeners to acknowledge *Gueniveve*'s transmissions and to offer assistance.

Distress Frequencies

Certain frequencies are designated as distress and calling frequencies, world-wide. There are also supplementary frequencies allocated as distress and calling frequencies for particular parts of the world. If you cannot raise anyone on a recognised distress frequency, you can use any other frequency you wish to try to gain assistance.

The recognised radiotelephone distress frequencies are:

Frequency/Channel	*Use*
156.80MHz/VHF CH 16	World-wide VHF distress and calling. Inshore distress channel; on-scene communications. Range 20 — 40 miles.
2182kHz/(no channel designator)	World-wide MF RT distress and calling channel. EPIRB (2182). Range up to around 150 miles for small vessels.
4125kHz	Supplementary to 2182 kHz, for use in Regions 1 and 2, South of latitude 15° North, including Mexico; and Region 3, South of latitude 25° North. Use when unable to gain assistance using 2182kHz.
6215.5kHz	As 4125kHz, but for use in Region 3 only, South of latitude 25° North.
8364kHz	For use by 'suitably equipped survival craft stations', world-wide.

If you are in contact with any station, ashore or afloat, on any frequency, when a distress situation arises — you need to consider whether it would be best to pass your initial message to that single station, on that frequency, before moving to one of the designated distress frequencies. If you only have time for a single transmission you will then be relying on your contact to use the Mayday Relay procedure. But you do know, at least, that *someone* will know that you are in trouble.

An alternative is to tell your contact, briefly, that you are in trouble and ask them to *'shift to CH16/2182kHz'* (or whatever) *'to receive my distress message'*.

Control of Distress Traffic
The following additional terms are used around the world, for the control of distress traffic.

Term	*Use*
'Seelonce Mayday'	Used *by the station in distress* to tell an interfering station to stop transmitting, as distress working is in progress on the frequency concerned.
'Seelonce Distress'	Used by another station involved in the distress working (eg, a CRS, the on-scene commander, or another vessel) to tell an interfering station to stop transmitting, as distress working is in progress on the frequency concerned.
'Prudonce'	Used by the station controlling distress working to indicate that there is still a distress situation, but that it is not deemed necessary to continue with complete silence on the frequency concerned. Communications can continue, calls should be kept to a minimum, and stations should expect a re-imposition of silence at any time.
'Seelonce Feenee'	Used by the station controlling distress working to indicate that distress working has now ended on the frequency concerned and that normal communications can resume.

The need to use *'Seelonce Mayday'* and *'Seelonce Distress'* is usually the result of a ship tuning-in to a distress and calling frequency and calling another station, without first establishing whether or not distress working is taking place.

If you switch on your receiver and tune-in to CH16 VHF or 2182kHz, in an area where there is normally a lot of radio traffic, to find that the frequency is quiet — do not assume that your luck is in for once! Wait some minutes to see whether anything else is going on.

Authority of The Master
The only people who can authorise a distress transmission from a particular vessel are the owners and the master. If the owner/master are unable to exercise that authority someone else would normally assume command and, therefore, the authority to transmit a distress message.

Misuse of the Distress Signal

Severe penalties can be imposed on anyone who misuses, deliberately or accidentally, the distress signal.

This includes such actions as the deliberate transmission of a false distress signal/message by a station not in distress, and the accidental transmission of distress signals from, for example, an Emergency Position Indicating Radio Beacon (EPIRB).

The penalties imposed vary from country to country but might include a fine, imprisonment and/or meeting the cost of any unnecessary search and rescue mission which was undertaken as a result of the false distress message.

The above should not prevent anyone in trouble from taking the necessary action to gain assistance but emphasises the importance of understanding the Distress and Urgency Rules, and being confident in their purpose and method of use.

When in doubt, make contact with the authorities ashore, explain your situation and take advice.

Summary

There are a large number of frequencies, services and facilities available to the mariner across various bands.

The maritime bands use various propagation methods, covering different distances, to pass messages from one station to another, thus making different bands suitable for different uses.

Frequencies are distributed to users by both international allocation and by national assignment. The international allocation ensures that mariners can cruise the world and can access similar services, with the same equipment, wherever they go.

In addition to agreeing what frequencies can be used for which purpose, the international community has also established standard procedures for exchanging information, breaking down potential communication barriers and ensuring the most effective use of equipment, services and facilities.

The chapters which follow explain the equipment, services, facilities and procedures for each of the maritime radio and satellite bands, and for alternative communication facilities open to seafarers around the world.

Chapter 3

Marine VHF Radio

Introduction

Marine VHF is likely to be the first 'two-way' radio system fitted on any vessel (in the USA, you won't get permission to fit MF/HF SSB *unless* you fit VHF first).

It is also the least expensive, most compact and most versatile short-range system available to mariners, meeting all the basic maritime communication needs at minimum operating cost, requiring little technical knowledge, and with greatest ease of installation.

This chapter describes the various types of VHF equipment available, the controls, features, capabilities and limitations of Marine VHF sets.

International channel allocations are described, along with the different, additional national channels you will need to consider if you intend to cruise around different countries.

Facilities for distress, urgency and safety messages are described, along with Port Operations, Ship Movement and Vessel Traffic Systems (VTS).

Inter-ship and Public Correspondence radiotelephone calls are covered, as is the use of private channels for business use, including yacht clubs and marinas.

Marine VHF Equipment

There are three main classes of ship's Marine VHF equipment — two for permanent installation and one temporary.

The two categories designed for permanent installation are (i) those which meet the most stringent specifications required of compulsory-fitted vessels, and (ii) those which will meet the everyday needs of most voluntary-fitted craft, including yachts, power-boats and fishing vessels.

The third type is the handheld set, for use in tenders and other small craft which do not have a permanent installation.

The operating capabilities of the two types of permanent installation have come closer together in recent years, and most owners of voluntary-fitted vessels can now usually find a set to meet all their requirements, without resorting to the most expensive equipment designed for compulsory-fitted vessels.

The three main items which distinguish those more expensive installations from others is:

* Full-duplex capability.
* An ability to support a number of 'phone points' onboard, from one main set.
* An ability to work on both main and stand-by power systems.

Sets made purely for voluntary-fitted vessels tend to work from the 12V/24V battery supply as their main (and only) power source. As many smaller craft only have battery supply the third point above would not therefore apply. It is a consideration, however, for larger yachts and other boats with an AC mains electrical system.

Larger yachts and boats might also have a requirement for radio access from different parts of the ship, in which case the second point would have to be considered.

A full-duplex capability does make radiotelephone calls more natural, as described later, but is not essential for most normal operations. Full duplex is required, however, for facsimile (fax) transmissions and other types of data communication, both of which are becoming more widespread — especially for the 'floating office'.

Before buying a VHF set, you need to consider all your future communication needs, to be sure you choose a set which will meet all your requirements. Some possible configurations are shown in Fig 3.1.

Use the table on page 70 to assess your own requirements — after reading the remainder of this chapter!

VHF Installation

How and where you install your VHF set will depend on the full range of additional facilities you might want to use. Two primary requirements which ought to be met, where possible, are:

* You should be able to operate the VHF from your conning position (or positions, if more than one).
* The antenna should be sighted at the highest point possible (for maximum range) and should preferably have an unobstructed, all-round view.

Antenna Sighting and Fixing

The second point above is normally solved by fitting the main antenna at the top of the tallest mast. If that is your *only* mast, then the VHF antenna will be in contention with others including Loran, Decca and television; wind direction indicators, anemometers etc.

In practice, sailing craft with more than one mast can normally sight their VHF antenna on any mast, with little loss of range. The most convenient can therefore be chosen, rather than the highest.

Other solutions include:

* Fitting a combined VHF antenna/wind direction indicator
* Fitting a multi-system, active receiving antenna (Fig 3.2)
* Fitting an antenna stub mast at deck level away from the mast, and re-siting antenna which are not height dependant (eg Decca, Loran, Omega, Sat-C) on the stub, so leaving the masthead free for the VHF antenna.

Sailing craft have the additional potential problem of fouling by, or of, sails and rigging — which will also restrict the options available.

Power boats, and others without tall masts, are left with the option of the antenna stub mast (or two) — or cabin top/side mounted brackets. Fortunately, there are a number of suitable brackets and antenna fixing options available to meet the needs of *all* types of craft, including power boats. (Fig 3.3)

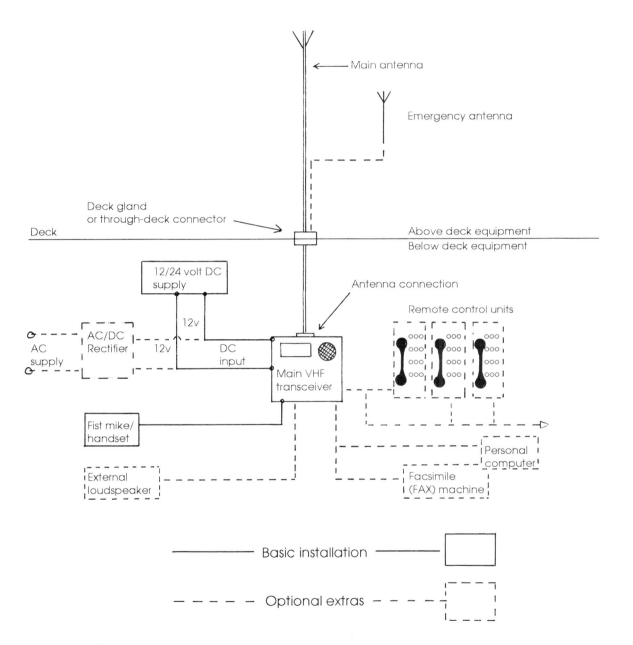

Fig 3.1 VHF radio installation

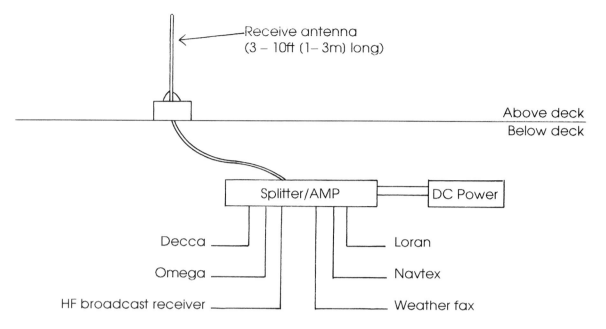

Active receive antennas are most effective when well away from the coastline. This is because, at 3 — 10 ft long, the antenna itself reacts strongly to local FM radio stations operating in the VHF bands and also other local short-range stations. These strong signals suppress the amplifier in the splitter box, so your 'wanted' signal does not get the level of amplification needed to activate the receiver.

Fig 3.2 Active Receive Antenna

Which Antenna?

It is best to discuss your requirement with the antenna supplier. Before you do, however, you need to consider a number of factors which might affect your choice and have your answers ready.

What environment do you expect your antenna to survive in? Are you a fair-weather sailor or one of those whose only pleasure comes from pushing small boats through rough seas?

If your antenna is going to be subject to heavy dousing and/or vibration, a rugged mounting (eg chrome) would be best. That means that you should be looking also for an antenna with a chrome, stainless-steel or nylon ferrule for fixing to the mount. Chrome, stainless-steel and nylon can be used together, but there are some zinc alloys which should not be used at sea where they will invariably seize-up.

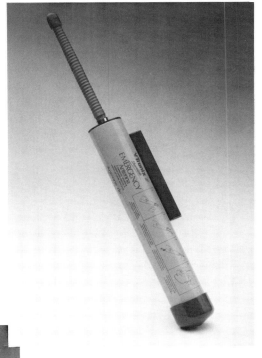

A combined VHF antenna and wind direction indicator.

An emergency antenna

A heliflex stub antenna — flexible, low-profile and approximately 150mm tall.

(Photos courtesy of VTronix)

Three different types of alternative VHF antenna

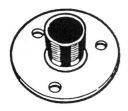

Simple deck mount
(Light duty)

Mast mount
(Light duty)

Rail mount

Ratchet mount
(For fitting on horizontal or vertical
surfaces)

Bracket mount
(For side-mounting or as an upper
support for an extension pole)

Illustrations by courtesy of V-Tronix Ltd

Fig 3.3 Antenna fixing brackets

How much 'gain' can you use?

Antenna gain is a measure of performance, whereby you can achieve better range according to the gain of the antenna.

Gain is not amplification — it is not a way of *increasing* power, but of concentrating the available energy (transmitter power) into a flatter doughnut beam, to give you more range for the same transmitter output. (Fig 3.4)

Antenna gain is normally stated in manufacturer's literature and is expressed as 'decibels', relevant to a standard. Three decibels (3dB) means 'twice as much' or 'double'; 'minus 3dB' would mean 'half as much'. The two common standards are the '*isotropic radiator*' and the '*dipole*'.

An isotropic radiator is a theoretical antenna which radiates energy equally in all directions. Such a perfect radiator does not exist, but does provide a useful measure to judge the performance of others.

A VHF antenna which provides 'unity' gain, will give the same horizontal line of sight range as a dipole.

An antenna offering '3dBi' is concentrating the available energy so that twice as much power is being radiated on the horizontal plane, than from an isotropic radiator. A 'half wave dipole' antenna has a gain of between 2dBi and 3dBi.

An antenna which offers '3dBd', will radiate twice the power of a halfwave dipole (hence 'd' rather than 'i', after '3dB'). This is nearly, but not quite the same as a 6dBi antenna.

The extra gain achieved by the 6dBi (or greater) antenna is won at a price. The antenna construction causes the transmitted power to be concentrated into a narrower beam, which can cause problems for vessels which are inclined to heel over a lot (pun not intended) (Fig 3.4ii).

Where reasonable height is available to sight the antenna, the dipole or the 3dBi antenna is usually appropriate — depending on the angle you expect to spend much of your time at!

For power boats and others with limited height, the extra gain from 3dBd or a 6dBi antenna can help compensate for range limitations through lack of height. Power boats are also more stable than sailing craft (ie, more likely to maintain an even keel), so the narrower beam-width should not normally be a problem.

Gaining an extra few feet is much more important for power boats than for those with a proper mast — as the extra range gained is much more significant where there is no really 'high' antenna (try some comparisons on the nomograph in Fig 2.2).

Aesthetics

It is worth considering the appearance of your antenna — particularly if you intend to fit a range of equipment which also require additional antennas.

Is your choice of VHF antenna going to be the odd one out? If you are going to have a number of antennas onboard, and it is possible to get them all looking the same (eg, all stainless steel or all white fibreglass) — *without* compromising on performance or durability — then the time to make that choice is before you spend any money on your first antenna, *especially* if it means changing mounts or brackets also!

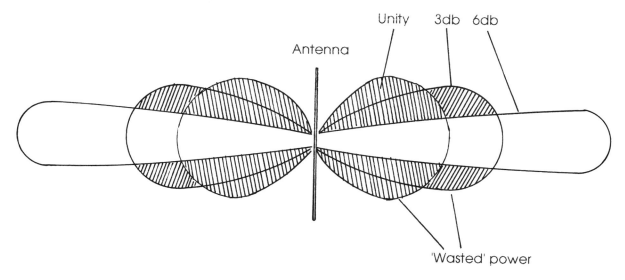

(i) The 'gain' is a way of concentrating the available energy into a narrower beam. As we go further from the transmit antenna therefore, the signal stays stronger for longer. The high gain also improves the receive antenna capabilities — allowing weaker signals to be captured.

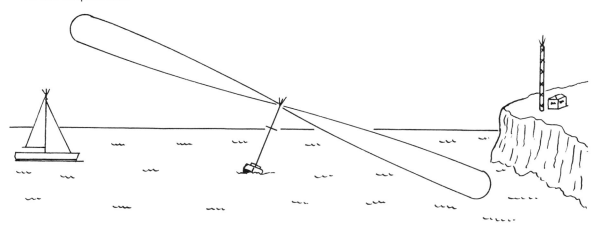

(ii) A high gain antenna could be a disadvantage in some circumstances, especially where a vessel is inclined to heel. The concentrated beam could result in failing to make contact with another station who may otherwise have been in reasonable radio range.

Fig 3.4 Antenna Gain

Antenna Cabling and Connectors

The antenna system, including cable and connectors, must present a specified level of *impedance* to your set. All modern fixed Marine VHF sets look for 50 ohms impedance from an antenna system, and this will be stated on the specification chart for each set.

For optimum performance the antenna itself should be rated at 50 ohms, as should all cabling and connectors. (Antenna, cabling and connector impedance is not additive.)

For convenience, waterproof *solderless* connections are best for the cable-to-antenna joint and for any thru-deck connection — provided that the connector you use is designed to be solderless in a marine environment. All cable-connector joints themselves must be good soldered joints, however, or corrosion will set in and performance will quickly deteriorate.

The normal cable for use with Marine VHF on smaller craft is designated as 'RG58CU'. This is a co-axial cable about 5mm (just under quarter-of-an-inch) in diameter, with a multi-strand, tinned copper wire inner core and an outer screen, also of copper. (See Fig 3.5) (The cable resembles that used for television antenna.)

The inner core carries your transmitted signal from set to antenna, and any received signal from antenna back to radio set. The screen acts to stop your transmitted signal from interfering with other radio/electronic equipment aboard — and to reduce on-board electrical interference to received signals.

RG58CU cable is rated at 50 ohms and is good for cable runs up to around 60ft/20m. At this length of cable run you will be losing half your power before reaching the antenna. At 25 watts transmitter output, more than enough power is left to ensure your still achieve good range. The limitations of long cable runs when attempting to use low power (1 watt), will be obvious.

For cable runs longer than 60ft/20m, the cable to use is RG213CU. Twice as thick as RG58, RG213 has the same nominal impedance, but a much lower (50% less) loss level.

The problem with RG213 is its thickness. At twice the diameter of RG58, it takes up more of that valuable space in narrow cable runs, does not readily turn corners, nor is it easy to feed through holes in awkward, restricted places.

Use of RG213 cable will, however, allow maximum output, even at low power.

Watertight, thru-deck, 'N-type' solderless plug-and-socket connectors are preferable to running a single piece of cable through a deck-gland and directly to the radio set. This makes replacement of damaged cable easier and, on sail boats, allows a deck-mounted mast to be removed with minimum additional effort.

Where the mast is stepped through the deck and fixed below, the cable can pass through the deck with the mast and can be connected directly to the set using the normal UHF/PL259 connector.

You should not be tempted to use TV cable with your Marine VHF set. TV cable might look the same as RG58, but its impedance rating is 75 ohms, not 50. Using the wrong cable will not only reduce the power reaching your antenna, it could result in malfunction of, or damage to, your set.

Some hand-held sets can be connected to a fitted 50 ohm main antenna system using a suitable connector. Check for connector compatibility with your fixed antenna system before buying a hand-held set if you want to use such a facility.

Co-axial Cable

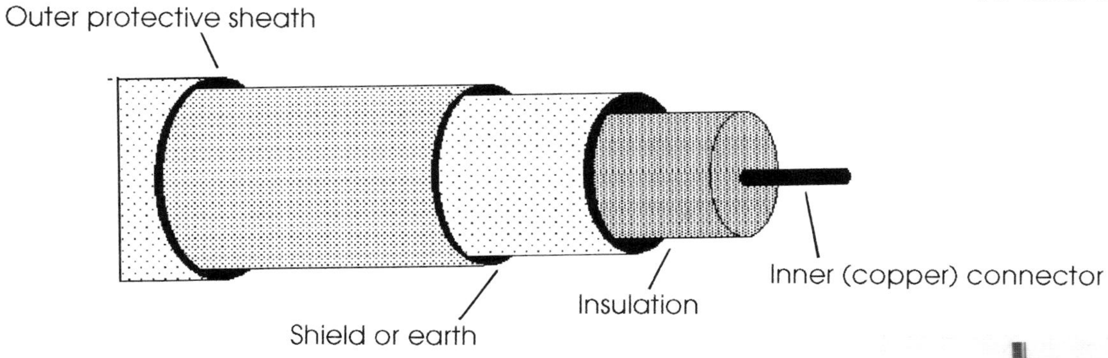

Outer protective sheath

Inner (copper) connector

Insulation

Shield or earth

N-Type Thru-deck Connector

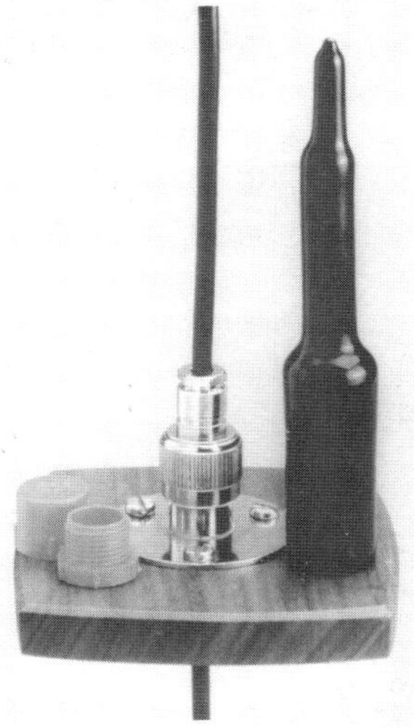

Photo Courtesy of VTronix

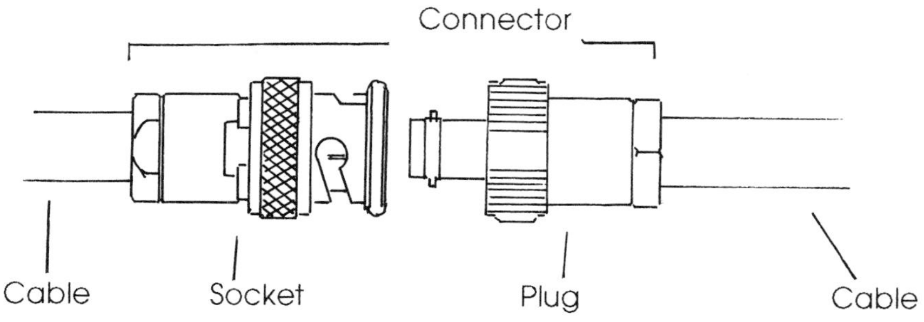

Connector

BNC Connector

Cable   Socket   Plug   Cable

Fig 3.5 Cable and Connectors

Who Can Install?

Unlike some other types of radio equipment, you do not need to hold any particularly technical qualification to carry out the physical installation of Marine VHF on vessels registered in the UK or the USA. (The position is believed to be the same for most other countries).

The vessel *does* need a Ship Radio Licence for Marine VHF equipment before it can be installed and operated, regardless of country of registration. The equipment to be installed must also be 'type-approved' by the country of registration (see Appendix Q for licensing authorities, etc).

You cannot carry out internal adjustments to the equipment, nor over-the-air tests, unless you hold an appropriate radio operator's qualification (See Chapter 10 for licensing and operators qualifications).

Power Supplies

Most Marine VHF sets will operate from a 12 volt or a 24 volt DC (battery) supply. If it is a 12 volt set and you have a 24V DC supply, you will need a 24V/12V DC converter. If your mains power is AC, you need to use an AC/DC rectifier. (Fig 3.1)

Marine VHF equipment draws very little current when on receive, but drains considerably more on transmit. As with other onboard electronics, a reliable power system is essential for continued operation.

Even on transmit, the current drain for Marine VHF sets, at around 5 amps, is relatively small. When adding *any* new electronic equipment to your vessel, however, you need to consider the cumulative effect of *all* the equipment connected to your power supply, to ensure that supply system will continue to meet all your needs.

Fitting the Set

The two things to consider when fitting the set itself are (i) where it is to be located, and (ii) whether to flush-mount, or bracket-mount.

VHF radio equipment is normally supplied with a bracket for fixing the set to the deckhead, bulkhead or to a bench top. This is the easiest type of fixing — but make sure any screws going into the deckhead are not coming back out somewhere else!

When flush-mounting, you must leave enough extra cable (antenna and power) behind the set to allow it to be pulled out for servicing when necessary.

Finally, if you are fixing your VHF set in an exposed position and have decided on a 'splash-proof' set — make sure the microphone/hand-set is *also* waterproof. If not, the lot might as well go below.

If the set does have to be fitted below-deck, an external (waterproof) loudspeaker will at least let you know what is going on around you.

Radio Frequency Interference (RFI)

Another location factor to consider is radio/electrical interference — to or by — the set.

As mentioned in Chapter 1, radio waves are electromagnetic. Your VHF set is therefore a neat little generator of electromagnetic energy. Fix it (or a remote loudspeaker) too close to a magnetic compass — and it will cause that compass to tell you lies.

Your VHF should be sited at least 18 inches (50cm) away from a magnetic compass or any other electronic position-indicating device. Heavy duty remote 'speakers should be even further away — so check the manufacturer's instructions before fitting.

When you first install your set, get someone to check the compass and any other electronic equipment, while you test the VHF to make sure you are not affecting its performance.

The VHF, like other electronic devices, can also *suffer* from interference, particularly from alternators, electricity generators and electronic ignition systems on outboard motors. You need to protect your VHF set from electrical interference.

Where electrical interference is a problem, your first line of defence is to make sure your VHF set is properly earthed. If you still have problems, then shielding or changes to the electrical installations causing the trouble are required.

When you have fitted the set, connected it to the antenna and are neither interfering with, nor being affected by other equipment — you can start twiddling the knobs!

Marine VHF Controls

Marine VHF sets have particular controls and switches which are available on all manufacturer's products. These include Power On/Off Switch, Channel Selector, High/Low Power Switch, Squelch and Quick Channel 16 selector.

Additional controls/features might include Dual-Watch, Selcall, Scan and Hailer/Foghorn.

The user control functions are as follows:

The *Power On/Off Switch* completes the connection to the vessels power supply, making the set live.

The *Power High/Low Switch* changes the *output* power of the transmitter from 25 Watts (the maximum allowed on Marine VHF channels) to 1 Watt. Using low power can reduce the battery drain to around one-quarter of that drawn on high power and should be the chosen setting for inter-ship, port operations and marina communications wherever possible.

High power is needed for radiotelephone calls to help the station connecting you to the telephone network to provide a solid link.

The *Channel Selector Switch* will either be a rotary control, a 'channel search' button or a numerical keypad. Most modern sets display the selected channel number on a digital readout.

The *Channel 16 Selector* is a one-touch push-button/switch selector, which will put your set straight onto Channel 16 (the International distress and calling channel), in case of emergency. It should also automatically put you on high power. (This is a statutory requirement for equipment approval in most countries).

The *Volume Control* is an audio stage control, changing the level of output from the receiver to the loudspeaker or handset. (Some remote loudspeakers will have a separate speaker control).

The *Squelch Control* is a convenience which will help stop you going mad when no proper signals are being received. Without squelch, a VHF set will produce continuous 'white noise' or 'mush', guaranteed to test the patience of the most reasonable of people. The squelch control suppresses that mush.

The squelch control will be a rotary knob, or a twin 'up/down' push-button arrangement. To set the squelch, proceed as follows:

* Turn the volume and squelch controls to their lowest level so that no sound is heard from the speaker/handset.
* Turn the volume control up to around three-quarters, when you should be hearing loud mush from the set.
* Slowly bring the squelch control up, until the mush just disappears.

When you hear your first signal you can then adjust the volume control, if required.

If the squelch control is set too high, it will stop weaker signals getting through to your speaker/handset — which could become embarrassing, dangerous, or both.

The *PTT Switch* is the one which can most confuse new users. Most people remember to push it to talk, but some (who me?) forget to release it to listen!

With a *full-duplex* set working on a two-frequency channel (eg, a ship-shore radiotelephone channel), the PTT switch can remain pressed throughout the call, making for a more natural conversation — including spontaneous interruptions! As most small craft sets are *semi*-duplex however, it is necessary to release the PTT switch to hear the shore party talking.

The result is that shore users *also* have to understand the basic operating procedures and, like the ship operator, need to say 'over' when they finish talking and want a response from the ship.

On single frequency channels like the inter-ship and main port operation channels, the PTT switch needs to be released when listening, even with full-duplex sets. That is because the receiver input, at the same frequency to the transmitter output, is automatically cut-off when transmitting to avoid damage to your set.

Additional Features

Dual-watch

Most sets now offer '*Dual-watch*'. This allows you to monitor a particular channel (eg, an inter-ship, or a Port Operations channel) without ceasing watch entirely on Channel 16. When Dual-watch is activated, the set will maintain watch on your chosen channel and will 'flip over' to Channel 16, briefly, every few seconds. If there is a signal on Channel 16, the set will remain on that channel so you know what is going on.

If there is no signal on Channel 16, the break in reception on the other channel is usually too short to disrupt communications — you receive enough of what is being said to avoid the need to ask for a repetition.

Unfortunately, some parts of the coastline in many countries are now so busy that Channel 16 congestion has reached barely tolerable proportions. Using Dual-watch in such a location would keep your set on Channel 16 almost continuously, which is not what Dual-watch was intended to do.

In such circumstances users often have to forget Dual-watch and stick to the other desired channel — changing over manually, on occasion, to check whether distress or other emergency working is under way.

If you have no good reason to maintain watch on another channel, then CH16 should get your ear, continuously, when at sea.

Selcall

Selcall (Selective Calling) is another facility provided with some sets. When applying for your Ship Radio Licence you can ask for a Selcall number to be allocated, in addition to the radio call-sign. Ship's Selcall numbers are five-figure numbers; Coast Radio Stations (CRS) are allocated four-figure numbers.

With your Selcall number programmed into your set, you will be alerted if the tones representing that number are received by your set. (Selcall signals are broadcast on Channel 16).

Some sets have the ability to display the Selcall number for the *calling* station, so you do not have to guess who it was that set off your alarm!

Selcall and Digital Selective Calling (DSC)

Selcall is the recognised name for the 'Sequential Single-Frequency Code System' — or 'SSFC'. SSFC codes are used to call ships (or other stations) on CH16, or on 2170.5kHz MF. The five digit ships Selcall code is transmitted as five musical tones over the air.

DSC uses different codes. Brought in to meet the requirements of distress calling as part of the Global Maritime Distress and Safety System (GMDSS — See Chapter 9) — DSC uses the nine-digit Maritime Mobile Service Identifier 'MMSI'.

Ships can ask for either or both to be allocated when applying for a licence or as a change to the existing ships licence.

Scanning

Scanning of a number of channels is also now a feature of many sets — either scanning all channels or scanning a limited number of channels selected by the user. Scanning is a bit like Dual-watch, but covering two (or more) channels, with Channel 16 not necessarily included in your selection.

Hailer

Some sets now also have a *hailer* option. Using an above-deck loudspeaker device, you can switch to 'hailer and use the facility to talk to people on deck (big boats!), onshore, or on other craft beyond normal talking distance. The loudspeaker also acts as a voice receiver, passing received speech back to your handset or built-in speaker. Such sets may also have an additional 'foghorn' facility (making a noise like a large strangled duck).

Marine VHF Channels

There are 55 International Marine VHF channels that can be used for various purposes when operating a Marine VHF transceiver under the terms of a Ship Radio Licence.

The general agreement on which channels can be used for particular purposes, (together with some local differences for Canada, the USA and the UK), are listed in Appendix D.

The main uses are (i) Safety of Life at Sea (SOLAS); (ii) Inter-ship; (iii) Port, Harbour and Ship Movement; and (iv) the Public Correspondence Telephone Service.

SOLAS Services

The most important channel in the VHF band is CH16. This is the International Distress and Calling channel and is probably the one channel which gets most use — and most abuse — around the world.

CH16 listening watch is maintained by ships and Coast Stations waiting for calls from other craft and stations. In this way, if a distress call goes out there will probably be someone listening. But of course, with the short-range nature of VHF, any distress call going out in thinly populated waters might well go unheard.

Coastguard Stations will normally control inshore casualty working on CH16, whilst all vessels involved in a distress incident anywhere will use CH16 for on-scene communications.

Broadcast and Safety Services — Canada and the USA

But there is more to SOLAS than casualty working. You might never be involved in a distress case yourself, but there are a number of safety services which you might want to take advantage of, every time you cast-off.

Weather forecasts are the most obvious example. In Canada and the USA, specific VHF channels are dedicated to the broadcast of weather reports. Known as 'WX' channels, ('WX' is the Morse code abbreviation for 'weather') there are seven of them altogether — covering the Eastern and Western seaboards, the Gulf Coast, and the St Lawrence and Great Lakes waters. (See Appendix D.)

The seven WX channels are used to broadcast weather forecasts continuously, 24 hours daily. (In Canada, the WX channels also broadcast navigational information.)

Maritime Safety Information, including navigation warnings, are also broadcast on VHF. In the USA these broadcasts are sent on Channel 22A, after an announcement on CH16.

Channel 22A is actually the ship-transmit frequency of international duplex Channel 22. Foreign vessels visiting North American waters really need to have this receive frequency fitted onto a spare 'private' channel on the VHF set, unless they have fitted a set with a 'USA/INT' facility. (USA/International sets can be switched from international frequency configuration, to US/Canada configuration, at the flick of a switch.)

US and Canadian vessels planning international voyages also need the USA/Int facility if they are to get the most of their VHF when cruising abroad.

Broadcast and Safety Services — Europe, Australia and New Zealand

The 'continuous weather' broadcasts are unique to North American waters. Anywhere else in the world, and you need to clock-watch.

The standard for weather forecasts around the world is for twice-daily broadcasts — typically, early morning and mid evening.

The international norm for gale and storm warnings is for an initial broadcast on receipt by the Coast Station, followed by repetitions at (scheduled) four-hourly intervals.

Port, Harbour and Ship Movement Services

Port Control and Harbour Masters tend to be allocated a simplex Port Operations channel as their primary channel — with CH11, CH12, CH13 and CH14 being the most common primary channels all around the world.

Using simplex channels for port operations allows other people to listen in and hear both sides of a conversation, thus promoting safety without excess calls. It also lets the Harbour Master work from the office or from a (pilot) launch on the same frequency.

Busy commercial harbours are allocated additional VHF channels — both simplex and duplex, according to their needs. When complicated or lengthy movements are under way, they can put the vessel(s) concerned onto one of these additional channels, leaving the primary channel free for establishing communication with new arrivals (or departures).

Vessel Traffic Services (VTS)

Busy or enclosed waters (like the Strait of Dover, the approaches to New York and the St Lawrence seaway), operate Vessel Traffic Services as an aid to safety of navigation. Commercial vessels have to report in to these services and have their movements controlled, a bit like the air-traffic control at airports.

Leisure craft should listen into the VTS frequencies to hear what is going on and to anticipate the movements of larger vessels.

Calling Procedures

Calling a harbour authority in smaller ports is generally not a problem if they are open to service and you know what procedure they operate.

Most small harbours will have only one operating channel and may, or may not, keep a Channel 16 watch. You need to know the working channel, whether it is a 'direct call' or CH16-first port and during what hours the radio is manned.

Some harbours only maintain radio watch for period of time either side of high water — if nobody can get in or out, there is no need to control traffic!

Leisure craft rarely need to call port operations services in the large ports. Neither are 'unnecessary' calls welcomed, even although your radio licence does cover you for most of the channels used by harbour authorities and vessel traffic schemes. Even if your calls are not particularly welcome, your ear will certainly be! You must know what channel is used for which purpose in the larger ports if you are to know what activity is going on around you.

For example: Port operations for Southampton Water and its approaches.
The area covers a combination of pleasure craft, merchant shipping and warship movements and is therefore one of the biggest VHF users in Europe. A number of VHF channels are used by Southampton Vessel Traffic Services and the adjacent Port Authorities to manage the waterways.

Channel 12 is, not uncommonly, the primary port operations channel. Southampton operates five other simplex channels, three duplex channels and two private channels. There is so much radio traffic that CH12 has been set aside as a *calling* channel, from which vessels are asked to move to one of the other channels to exchange all but the briefest of messages.

Southampton VTS also uses CH12 to broadcast details of large vessel movements, every two hours, during the summer weekends and bank holidays. (See Appendix D which list the channels operated by port authorities.) Your initial call should indicate which channels you can use. (Apparently, it is not uncommon for a yacht with older VHF equipment to agree to shift to a particular channel, only to find that the channel is not fitted to their set!)

Marina Channels
Different countries have approached the need for a marina channel in different ways.

In Canada and the USA, the simplex CH68 is the most popular channel available for 'inter-ship and ship-shore' communications and is used by many marinas. CH71 (Canada) and CH78 (USA) are also used by marinas. Some commercial marinas in the USA are allocated CH9 as their primary channel.

In the UK the duplex CH80 is now the primary marina channel. This follows an abortive attempt to cater for this need by allocating 'Channel M' (private Channel 37) for British craft and 'M2' for others. CH-M is now an overload channel for CH80 and M2 is used for special events (race control, etc).

Inter-ship Channels
CH6 is a primary inter-ship channel all around the world. The emphasis on how it can be used however, differs in different countries.

Yachtsmen from the UK who are used to CH6 as a 'chat' channel might find themselves in trouble when visiting USA waters if they are not careful. In the USA, CH6 is specified as an 'Inter-ship Safety' channel and it must not be used for purposes other than safety. CH72 is the only inter-ship channel for use either side of the pond which does not have additional local restrictions.

Regardless of 'local' restrictions, inter-ship VHF channels are only legally available for use *on ship's business* and were never intended for social chit-chat. If you are in range of the authorities anywhere, then you need to think before you use (and particularly before you abuse!) the VHF facility. We will all stretch things now and again, but a practice that *none* of us should support is the abuse of CH16.

CH16 really is a lifeline for people who suddenly find themselves in trouble, possibly with only one chance to get a call for help out. Yet it is continuously over-used by people making lengthy and frequent calls when trying to contact other stations. There are also still too many people who consider it 'OK' to pass 'short' inter-ship messages — so:

1 Stick to inter-ship channels for regular inter-ship communications.

2 Use CH16 as a calling frequency *only*, not one to pass even the briefest of messages.

3 Avoid lengthy calls or too-frequent calls by following sensible procedures eg
    * Listen for a few minutes when first switching on, to make sure the channel is free from distress working.
    * Keep calls brief by using names/call-signs once only. If conditions require more than that, three is the maximum allowed.
    * If you do not get a reply to your first call, wait three minutes before making a repeat call and, if still no reply, wait another three minutes. If no reply to your *third* call — wait ten minutes before repeating the process.
    (Or consider what other channel your friend could be monitoring?)

On Inter-ship and other frequencies, all transmissions must be identified — even after you have established communications. This is best achieved by starting each transmission with 'this is ...... (vessel's name/call-sign).

Public Correspondence

All PC channels are duplex channels and are there to connect you into the shore telephone system of the country called. In the USA, the connection is made by the 'Marine Operator'. In the UK it is made by a BT (British Telecom) Radio Officer. The stations making the connection are officially designated as Coast Radio Stations (CRS) — whereas all other shore-based stations are known simply as Coast Stations.

As mentioned earlier, you should normally use full power when making a call into the telephone network. This is because the CRS equipment performs best with a good solid signal. Adjacent CRS's are allocated different channels to minimise interference, so using full power should not normally cause problems to others.

During the summer months VHF channels are occasionally affected by 'lift' conditions, when signals are carried much further than normal. This is the one occasion when VHF channels do not provide clear, consistent communications and when CRSs which use the same frequency and which are normally well separated geographically, find their channels disrupted by distant vessels. There is little you can do about it when it happens, but persevere and, again, keep your calls to the minimum required to get your message across.

Commercial/Business Radio

Special 'Base' sets are available for Marinas, Port Operators etc, who gain a licence to operate on an international channel or who are allocated a private (business) channel. The licence for a marina would not normally allow them to operate on any other channel than their allocated working channel(s). Not even on Channel 16. These operators might also be licensed to operate at lower than the 25 watts maximum power allowed for vessels.

Additionally, erecting an antenna ashore might require you to get permission from your local county officials — don't assume that having a licence to operate a Coast Station also allows you to erect whatever antenna you like, at any height, attached to any building — check it out first!

Further Information

The 'Admiralty List of Radio Signals' (ALRS) is considered by many to be the definitive document for world-wide lists of frequencies, operating times etc. for maritime mobile services on all frequencies. There are also national publications (eg 'Radio Aids to Marine Navigation' [Canada]) which cover maritime radio services — also, various yachtsmen's almanacs and other privately published references list maritime radio services to varying degrees.

Summary

If you only ever fit one type of marine radio — it has to be VHF.

Marine VHF is the *only* system which provides *direct* access to distress and safety facilities, port operators, pilot vessels, vessel traffic services, marinas, other craft and to the public telephone network.

VHF is the least expensive two-way marine radio to buy and install and is probably the only one which, in one form or another, can be used on any vessel.

Except for your radiotelephone calls, all the services on Marine VHF are free. It costs you nothing to listen to WX broadcasts, ship movements or to contact the marina or other vessels.

Even those radiotelephone calls which, minute for minute are expensive when compared with some other services, can be cheap when you consider the fact that there is no 'quarterly rental' or other standing charge to pay. Especially if you only make the occasional call.

The main limitations of VHF are (i) it is short-range; (ii) the congestion on Channel 16, the public correspondence and the inter-ship channels in busy areas; and (iii) the fact that inter-ship channels cannot legally be used for 'social' calls ('chit-chat').

Some of those limitations can be tackled by adding to your VHF, and some by using alternative systems — and that is what the next chapter is all about.

Assessing Your VHF Requirements

Function	Essential	Desirable	Product 1 Yes Price		Product 2 Yes Price	
All International Channels						
USA Channels						
10 x W x Channels						
Dual-watch						
Channel Scan						
Selcall						
Hailer						
Remote Loudspeaker						
Foghorn						
Main/Standby Power Selection						
Full Duplex						
Weather Resistant						
Splashproof						
Multi-phone points						

Note that quick Channel 16 selection, squelch, volume control, channel selector and High/Low power switch are not listed as they should be common to every set.

Chapter 4

Marine VHF Additions and Alternatives

Introduction
Chapter three explained the many benefits of Marine VHF and the limitations of the standard international set-up.

Those limitations were (i) congestion on CH16, which could cause a distress call to be missed; (ii) the ban on 'social chit-chat' on inter-ship channels; (iii) the limited number of public correspondence channels and the fact that radiotelephone calls are connected by a 'marine operator'; and (iv) the limited range of VHF.

This chapter describes ways you can get around those limitations — with emphasis still on short-range systems.

Gaining attention on VHF when CH16 is chock-a-block is covered.

Citizens Band radio — the system provided for 'social' calls is described.

A number of countries now offer an automated radiotelephone service on Marine VHF channels, using an 'add-on' unit for your VHF set. The various automatic VHF services are explained.

Automated *alternatives* to Marine VHF are also covered, including the 'Automated Maritime Telecommunications System' implemented in the USA, and cellular radio systems and their likely future role in maritime communications.

Channel 16 Congestion
One of the *qualities* of VHF is its clarity of calls. This is achieved because of a phenomenon known as 'capture' effect. But there is a down-side to the word capture, as explained below.

Capture effect (Fig 4.1) is a function of frequency modulation. In FM mode, only the strongest signal is reproduced by a receiving station and, once locked-on to a signal, the receiver will reject, completely, any weaker signals. You do not hear the weaker signals 'in the background'.

This is why FM radio is so clear — you either get a good signal or nothing at all.

In Marine VHF, and particularly in busy waters, capture effect could mean that a distress call might well go unheard. Especially if the casualty's main antenna came down with the mast or if a low-power, hand-held set is being used.

What happens is that VHF receivers in range of the distress call are 'captured' by stronger signals at the expense of the weaker signal from the casualty.

If you are having trouble getting through, your VHF alternatives are (i) switch to another channel, or (ii) use Digital Selective Calling (DSC). Or you can fit an Electronic Position Indicating Radio Beacon (EPIRB) for such an eventuality.

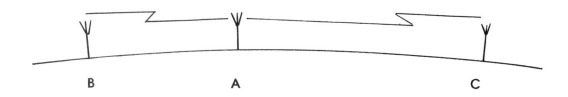

B and C are putting out the same power but if both are transmitting at the same time, only B will be heard by A.

Even if C transmits first and is initially heard by A — when B starts transmitting with the same power output as C, B will overpower C on A's radio.

The fact that B or C might be out of range of each other, so that B never heard C in the first place, simply compounds the problem.

Fig 4.1 Capture Effect

Change Channels?

You need to consider carefully before moving off CH16 and then only switch to a channel where you are *more likely to be heard*. If CH16 is congested, it is also likely that the inter-ship channels are busy.

But are you so far from shore that only other vessels will be able to hear you? Are you in range of a Coastguard Station, a CRS Marine Operator or a Port Operations service? This is where it pays to know what stations operate in your location, whether in your own home waters or wherever else you might be visiting.

Digital Selective Calling?

DSC is the digital signalling method for use on VHF and MF/HF under the Global Maritime Distress and Safety System (GMDSS — see Chapter 9).

DSC is used on VHF CH70 for inter-ship, ship-to-shore and shore-to-ship alerting but many countries do not monitor CH70 for DSC calls and only big ships are obligated to maintain a watch when at sea.

Again — you need to know what facilities are provided in your area before relying on VHF DSC to gain attention.

Electronic Position Indicating Radio Beacons

EPIRB's have been with us for a number of years and there is now quite a range of units on offer. The right choice of EPIRB will not only give you an alternative to Channel 16 in busy waters, but will allow you to alert shore authorities when at sea and out of VHF range —

anywhere in the world.

There are three main things you need to consider when choosing an EPIRB and those are:

1 On what frequency (or frequencies) do you want it to operate?

2 Do you want an automatically activating, float-free unit or one that is manually operated?

3 Do you want a unit which is complete in itself, or one which will include a position report from your GPS/Decca unit?

Earlier models of EPIRB worked on the civil and/or military aircraft distress frequencies of 121.5MHz and 243.0MHz, respectively. As aircraft listen on these frequencies *when clear of land*, these were a good option for taking out to sea. Operating on frequencies which use line-of-sight propagation, they can obtain considerable range skywards — and there is now also a satellite watch on 121.5MHz.

On picking up a signal from one of these beacons, the aircraft or satellite identifies the approximate location of the casualty (very widely approximate in some cases). Search and rescue aircraft go towards the area indicated and use a homing device to narrow down the location. Rescue vessels can then be directed to the approximate area to carry out a surface search. (Note: Uncorroborated 121.5/243.0 EPIRB signals may not trigger a SAR operation.)

As you can see, the title 'Position Indicating' is a bit of a misnomer. This type of EPIRB certainly sends out an alert beacon, but it does *not* indicate the position of the casualty. It takes direction finding equipment on SAR aircraft and ships to do that.

A more recent introduction is the 406MHz EPIRB, which operates within the 'Cospas-Sarsat' scheme for distress alerting and position fixing (also covered in Chapters 8, 9 and 10).

The same Cospas-Sarsat satellites which monitor 121.5MHz also monitor 406MHz but, in this case, they are able to pin-point the EPIRB to within a few miles — a vast improvement on any previous system.

The most recent entrant to the market is the 'L-Band' EPIRB, which works into the Inmarsat system (Chapter 7 and 9). L-Band EPIRB's can actually transmit your position, obtained from a Global Position System (GPS) unit, over the air. Some units have their own integral GPS receiver for this purpose.

EPIRB's offer not only an alternative to a crowded Channel 16 for distress alerting. They also give you a means of alerting the authorities when you are well out to sea — when you have Channel 16 completely to yourself — with nobody else within range to interfere with your VHF transmissions nor to pick them up when you really want them to.

A ship radio licence to cover VHF will normally also cover Radar and EPIRB's. You do need to register any such additions (see Chapter 10), but this is normally a formality.

EPIRB's are summarised in Fig 4.2.

Do not forget the non-radio options and flares in particular. Even where a distress call goes out on CH16, a visual indication is often necessary before the casualty can be picked out of the multitude of other craft which might be milling around the same area.

Operating Frequency	Remarks
121.5MHz	(i) Monitored by civil/commercial aircraft and Cospas-Sarsat satellites. (ii) SAR aircraft can home-in with DF equipment. (iii) A SAR operation will not be authorised solely because a 121.5/243.0MHz EPIRB has been heard.
243.0MHz	(i) Monitored by military aircraft. (ii) SAR aircraft can home-in with DF equipment.
406 MHz	Cospas-Sarsat satellites can pin-point position to within a few miles. (See Chapter 8 for implications of time lapse.)
L-Band	Continuous monitoring by Inmarsat satellites — these EPIRBS transmit your actual position as obtained from a GPS unit.

Note 1 All of the above can be supported by a SART (Search and Rescue Transponder), to allow surface ships to home-in on a casualty, once in the general area. (See Chapter 9)

Note 2 If you end-up in the water with a 'personal beacon' attached to your sleeve — one which operates on 121.5MHz — do not forget that the antenna must be above the surface of the water for efficient transmission. Also when you are in the water, the visual (and radio) horizon is minimal.

Fig 4.2 EPIRB Summary

Social talk and Citizens Band Radio

The only legal way to engage in social talk with your Marine VHF is through a radiotelephone call. As everything else on VHF is free, however, we tend to be shy about paying for a conversation between one boat and another. And many CRS's have only one working channel so, if you are both in range of the same VHF shore station you cannot both use that sole working channel at the same time.

The short-range alternative to Marine VHF which caters *legally* for direct, station-to-station social chatter is CB. It is fairly cheap and quite easy to install, and operating at 26 — 29MHz on the ground-wave only, has but a limited range. However, it does allow you to call other people direct and exchange your personal messages without charge.

There is also an 'emergency' channel on CB which might be monitored ashore and, then again, might not. CB sets will not be 'water resistant' nor 'splashproof' and are unlikely to tolerate excessive condensation.

Mariners should not seriously consider installing CB as an alternative to Marine VHF. Not only because of the more limited range, nor because of the lack of any official distress watch — but because CB will not give you access to the many broadcast services such as weather and navigation information, which might stop you getting into trouble in the first place. Also, a CB set can only be used in its licensed country as standards differ.

However, there is nothing to stop you gaining a licence and installing a CB set for use in your own national waters, either inter-ship or between vessel and shore, except for lack of space for the antenna, or that you could be using space which could be put to better use on another system.

Automatic Radiotelephone

There are three ways of automatically accessing the public telephone service by using maritime radio systems. One is via a satellite and the other two are automated terrestrial radio services. Both of the terrestrial services depend on where you are at the time — off the coast of where? In which waterway system?

The first is via an automated radiotelephone service over Marine VHF and/or MF/HF channels. Automation does not increase the number of PC channels available, but it does make for better use of air time.

On operator-connected calls, considerable air-time is lost while the vessel and shore operators exchange call and accounting details. Automated systems eliminate the need to pass vessel/call information manually and leave more channel time available for actual calls.

France was the first country to introduce an automatic Marine VHF radiotelephone service. The French service provides separate PC channels for its manual and for its automated service. French VHF stations also use direct-calling on the working channel for both manual and automated PC services, reserving CH16 for distress and safety traffic only.

Australia and New Zealand also operate automatic R/T services over Marine VHF channels — the Australian 'Auto-Seaphones' service and the New Zealand 'Sealink' service using a compatible tone-pad microphone so that users, once registered, can access the systems of either country. Your set will need adjusting to suit the new mike, so make sure it is first connected and tested by a marine technician.

Australia operates regular (manually connected) Seaphones and Auto-Seaphones on the same channel as the automatic service in some locations, and dedicated Auto-Seaphone channels in others.

The New Zealand Sealink service has replaced the manual service operated by Telecom Mobile Radio — the major provider in most areas, with Auckland Radio now being the only station in the Telecom Mobile Radio service still offering manual radiotelephone connection. There are other local NZ providers offering a manual VHF RT service.

In Europe and the USA, an automatic radiotelephone service based on an Italian model has been introduced.

The service is not restricted to VHF and promises to open up the maritime automatic radiotelephone service to world-wide access, as more countries fit the system.

In the UK, it is known as 'Autolink RT' and is operated on VHF, MF and HF R/T channels. The same 'Autolink' unit is used with VHF and SSB sets.

In the US, AT&T use this kit to provide their 'High Seas Direct' service on HF. And you can use the same unit to access automatic VHF, MF and HF services in a growing number of countries, world-wide (See Chapter 6, Fig 6.2).

One of the changes which automatic R/T has brought about is the need to pre-register before accessing the services of any particular country. The shipboard units used for AT&T's High Seas Direct and the UK Autolink RT service are supplied with their own unique identification number programmed in.

Personal Identification Numbers (PIN)

When you register for service through either country — or through any other country operating a compatible service — you will also be allocated a Personal Identification Number.

The unique unit number and your PIN are then registered on the service provider's control computer and each time you make a call, the automatic unit number and the PIN are compared. In this way, unauthorised use of your radio is curtailed (on the automatic RT service, at least). Additionally, you can ask for more than one PIN number to be allocated to your unit and you will then be billed separately for each PIN registered, with charges being allocated to the account indicated by the PIN used for any particular call. The multi-PIN facility allows you to separate business and private expenses, or expenses proper to different individuals onboard.

Like the Australia/New Zealand unit, your Autolink/High Seas Direct kit has to be fitted by a technician who will adjust the audio line input to your set (or sets) to match the output level from the automatic RT unit.

As with any new fitting, you need to go through the 'RFI' loop to check whether your new unit is interfering with any other electronic equipment, or whether the unit is picking-up spurious signals from other equipment.

Automatic Marine VHF makes better use of air time, but does not increase the total number of VHF Public Correspondence channels available to mariners. If you are a regular or heavy user of Short-range RT services, you might want to consider one of the other alternative systems for radiotelephone connection.

Automated Maritime Telephone Service

The US is the only nation so far to have introduced a totally separate, additional, automatic radiotelephone service exclusively for mariners. Known as the 'AMTS', the service uses VHF channels (in the 216 — 220MHz band), previously allocated to domestic television, to provide this unique service.

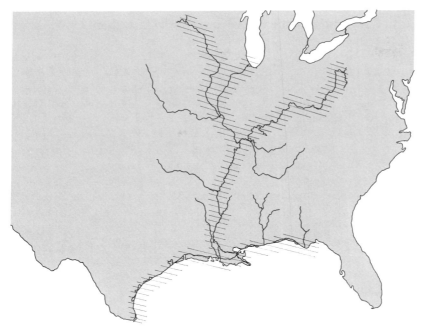

Fig 4.3 Watercom AMTS Coverage

There is at present a single licensed provider of AMTS in the USA and that is 'Waterway Communications System Inc', who use the registered name of 'Watercom'. The Watercom AMTS covers most of the Gulf Intra-Coastal Waterway (and adjacent coastal waters) from Brownsville, Texas, on the border with Mexico, round to Panama City on the Florida Panhandle. It also covers the Mississippi River from the Gulf to Minneapolis, the Illinois River from Chicago to the Mississippi and the Ohio River from Pittsburgh to the Mississippi. (Fig 4.3)

Watercom covers about 4,000 miles of inland waterways, with 55 shore stations — an average of one station every 70 miles. That means that stations need to achieve a range of around 35 miles in each direction if full coverage is to be maintained. Not excessive where reasonable antenna height is possible. The stations are monitored around the clock from a control centre at Jeffersenville, Indiana.

To date, the Watercom AMTS has been marketed primarily to the commercial sector, including towboats, cruise and passenger vessels. This has led to developments allowing individual billing for crew members and payment by various credit and calling cards for passengers. The commercial emphasis has also led to the development of facsimile and data transmission for the exchange of lengthy messages regarding cargo, etc.

But, like the normal maritime VHF service, there is no minimum monthly bill for using the system. Leisure craft who anticipate a reasonable usage, who frequent the rivers and waterways covered by the Watercom AMTS, and who find existing VHF Marine Operator services inadequate for their needs, should certainly consider the AMTS for their public correspondence radiotelephone calls.

The benefits for the 'floating office' from this system will be obvious. However, Watercom has no plans at present to extend coverage further down the Gulf Coast of Florida so, you need to consider where you will be spending most of your time before making any decision.

The Watercom AMTS is similar to cellular radio systems — with two exceptions. The first is that the AMTS is provided *exclusively* for mariners. Road vehicles cannot gain mobile access to the radio channels, except in the to-ship direction and through the public telephone network. Mariners therefore need not worry about being the poor relation in an AMTS. The second difference is that cellular radio systems operate in much smaller 'cells' than the AMTS, with stations typically only a mile or two apart. Users moving from the service area of one cell to another are 'handed-off' by one and automatically taken on by the other.

With AMTS stations being much further apart, and with river traffic being much slower than road traffic, 'hand-off' is not employed on the Watercom AMTS. Instead, ship-station location is monitored by the central control facility who arrange to re-route any to-ship calls from the public telephone network through the appropriate shore station. (Subscribers to the Watercom AMTS are provided with a standard 'NPA-NXX' telephone number, just like a shore telephone user).

Licensing of AMTS
The FCC has accepted that any vessel which is licensed for Marine VHF is also considered to be licensed for AMTS. As an AMTS is likely to be an addition to your Marine VHF you do not, therefore, need a new licence.

Cellular Radio Systems
Cellular radio systems have become increasing popular with leisure craft owners over recent years — in some countries more than others.

Countries with large unpopulated or thinly populated areas, like Canada, the USA and Australia, tend to have good cellular radio coverage around the major centres of population but patchy or irregular coastal or waterway coverage.

Other countries, particularly in Europe and Scandinavia, have excellent coverage of most of the coastline (the populated parts, that is).

The UK, with the exception of the West/North-West Coast of Scotland, has almost complete coastal coverage from two competing cellular service providers — and that is where some confusion can set in for potential users.

The two service providers, BT (British Telecom) with its 'Cellnet' service and Racal Decca with 'Vodaphone', are not the only people selling the cellular radio service. There are a large number of 'airtime retailers' in the UK who sell a 'package' to customers, sometimes with a giveaway price for the telephone installation, but with loaded call charges and/or a long-term contract.

It is yet another one of those cases which proves that there is no such thing as a free lunch or, in this case, a 'free' telephone! If you have considered all the maritime alternatives (see Fig 4.4) and still feel the need to subscribe to a cellular radio service, make sure you really are getting the most beneficial deal for your own circumstances.

There is no doubt that a good cellular radio installation will solve many of the problems of access to those crowded Marine VHF channels, but if you are only an occasional user of PC services, or if you are in an area of limited coverage, then you could be letting yourself in for more than you bargained for. Consider the up-front price of equipment and installation against length of contract and proposed air-time charges, for a number of systems and from more than one supplier. Sleep on it and then decide.

Problem	VHF Additions	Alternatives*
Distress alerting limited	DSC	EPIRB
Public Correspondence Telephone Inadequate	Automatic VHF Telephone Unit (Depending on country)	(i) AMTS(USA) (ii) Cellular Radio Systems
Limited Range of VHF Radio	Better antenna Arrangement	(i) MF/HF SSB (ii) Satcom

* Please note that 'alternatives' are alternative ways of tackling the problem stated — it is not suggested that any other system should be fitted without a VHF radio installation.

Fig 4.4 Additions/Alternatives to VHF

Pan European Cellular Radio

One of the bugbears of European sailors is that a cellular radio subscriber in one country cannot use their cellphone in another. For example, anyone leaving England for the Mediterranean will be out of UK cellular coverage before leaving the English Channel.

One answer for the mid-1990's is the new, Pan-European 'Group System Mobile' (GSM) — a cellular radio system which will allow the same mobile telephone equipment to access the new cellular service in each European country which 'joins the club'. Of course, there is nothing to say that the network development will provide as good coastal coverage as presently enjoyed by most Northern European countries so, once again, check the claimed and proposed service area and prices before you sign-up.

Licensing of mobile cellular radio transmitters is not a problem. Authorities tend *not* to require any licence at all. This is because the system works on fixed channels only (like Marine VHF) but, unlike Marine VHF, the operator cannot choose a channel on which to transmit. Cellular systems, like the AMTS, are fully automatic and controlled entirely by the host (shore) station. You do not even need a Marine VHF licence to operate a cellular set from onboard your vessel. Once again, however — the standard health warning. Please do not fit a cellular system *instead* of Marine VHF.

Cellular radio is designed to give you access to the public telephone network only and, even though this also allows you to dial-up the emergency services (Coastguard, police etc) it should not be seen as a *substitute* for Marine VHF, in its entirety.

Global Cellular Radio?

The development of cellular radio services to provide more global coverage is likely to become a key factor for mariners in the future, where the public correspondence telephone service is concerned. The Pan European System will allow individuals to subscribe in their 'home' country and have 'roaming' agreements with service providers in other countries and a compatible system is being installed in Australia. Similarly, the Advanced Mobile Phone Service and providers offering the American Digital Standard, will allow much more widespread 'roaming' than users in the USA enjoy with current systems. Future marine equipment is likely to provide both satellite and cellular radio access from the same unit — the equipment selecting the most appropriate (ie, cheapest) system depending where you are at any one time, subject to the 'roaming' agreements you have signed up for.

Combined satellite/cellular services are for the late 1990's and the 21st Century. Like today's cellular services, they will be for those people who require regular and ready access to the public telephone network. Those who only make the occasional telephone call will probably stick to the PC channels of the Marine VHF/SSB services, but will have to be prepared for a change towards more automation and less operator intervention as the service providers look to contain costs whilst retaining the available market.

Limited Range

Although we can optimise the range for our Marine VHF by using the most appropriate antenna, cabling and siting that antenna to our best advantage, the only real answer to the range limitations of VHF is to get into bigger things. Like Marine MF/HF SSB and/or satellite. As you progress up the ladder of marine radio you gain range but you lose other things.

Your MF/HF SSB set, for example, will provide you with (longer range) access to safety and weather broadcasts, inter-ship frequencies and public correspondence, but no direct access to Port Authorities.

Satellite systems provide weather and safety broadcasts, and public correspondence voice and data channels, but no direct, free, inter-ship nor port operations channels. Once again therefore, you are looking for an *additional* system, not a replacement for your Marine VHF.

Summary

Marine VHF is a low-cost, excellent, versatile and virtually indispensable piece of equipment. But it does have its limitations. Mariners are faced with various choices when trying to overcome some of those limitations, choices which add to the existing VHF installation or which provide some facilities using equipment which is separate from, and independent of, the VHF.

For distress alerting you can choose to add a DSC unit to your VHF; or you can obtain an EPIRB.

You can add an automatic radiotelephone unit or you can subscribe to a separate (short-range) service for your public telephone access.

In the USA, you can choose between the AMTS — the only separate short-range radiotelephone service *exclusively* for mariners, and cellular radio systems.

In Europe and, especially in the UK, you can choose between independent cellular service providers and a host of 'airtime retailers' all providing similar coverage and facilities. Or you can opt for the Pan-European system rather than be restricted to the service of any particular country.

The fact that Marine VHF operates over a limited range ('the short-range service') is a weakness, as well as its strength. Its strength is the variety of essential facilities it offers and which can be replicated over and over again, re-using the same channels, with limited geographical separation. Its weakness is that for blue-water sailors, you do not have to go far before you are out of range of land and of the various shore-based facilities you might want to use.

The additions and alternatives to VHF were summarised on Fig 4.4.

The next few chapters cover the options for deep-sea sailors, of medium and long-range radio and satellite services, the options which, with your VHF, provide the total spread of facilities and distance to allow you to keep in touch wherever you may be.

Chapter 5
Medium and Long-Range Radio

Marine SSB 'Side-band' Radio

Introduction
So you have your VHF and are considering fitting a satellite EPIRB for that ultimate emergency, inshore or on the high seas. You do intend blue-water cruising and want another communications facility — one which will allow you to maintain two-way contact with those back home, not to mention the ability to communicate in something *less* than a full-blown distress, like a medical emergency or other *'Pan Pan'* situation. You are considering installing a Marine SSB set.

This chapter explains the main services and facilities you can access with an MF/HF SSB radio, without any 'add-ons'. It covers the medium range (1.6—4MHz) radiotelephone band and the use of 2182kHz for distress and calling purposes. The long-range (high-frequency) radiotelephone bands are also described — where you can get service and how to make contact. Weather and navigational broadcasts on MF and HF are explained, as are frequency standard services and time signals.

Inter-ship communications on both MF and HF are covered and also long-range broadcast services like those of the BBC, the Voice of America, Radio Canada International and Radio Australia are introduced.

We start with a brief description of the types of installation available which you might care to fit and end with a description of the physical implications of installing Marine SSB.

Marine SSB — Types of Installation
There are two main types of SSB set suitable for installing on voluntary-fitted craft. The first is the one-piece transceiver with integral control panel (Fig 5.1a); the second has a separate transceiver unit and control panel (Fig 5.1b).

Whichever you fit, you will need an Antenna Matching Unit (AMU/ATU), a suitable antenna and an earth mat, or 'counterpoise'.

The type of radio you choose will depend on the size of your craft, where you want to control the set from and what services you are likely to use on a regular basis. You might also be influenced by the possible need to use the set in an emergency (distress or urgency) situation.

Medium and Long-Range Services
Maritime radio services in the MF and HF bands are referred to as medium-range and long-range services, respectively. The reason for this is that the MF services are provided by ground-wave coverage (see Chapter 2) with solid service from the coastline up to around 200 miles from the radio station concerned, whilst the High-frequency (long-range) service uses sky-wave to provide those more distant contacts.

82

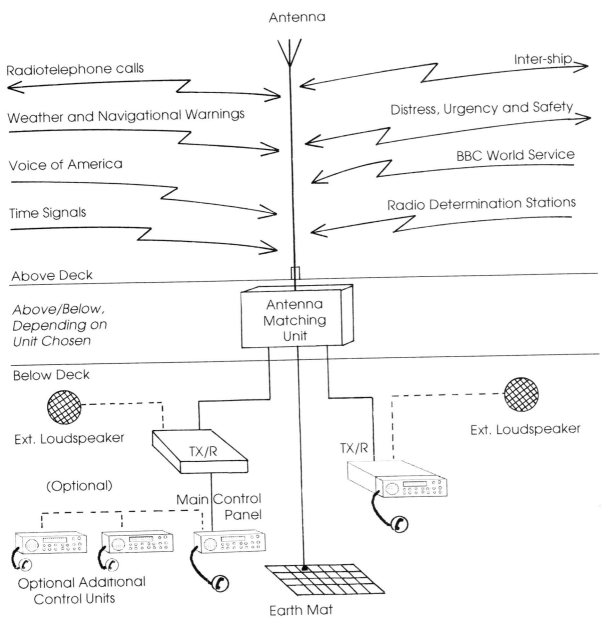

Antenna

Radiotelephone calls

Inter-ship

Weather and Navigational Warnings

Distress, Urgency and Safety

Voice of America

BBC World Service

Time Signals

Radio Determination Stations

Above Deck

*Above/Below,
Depending on
Unit Chosen*

Antenna
Matching
Unit

Below Deck

Ext. Loudspeaker

Ext. Loudspeaker

TX/R

TX/R

(Optional)

Main Control
Panel

Optional Additional
Control Units

Earth Mat

5.1b Control Units Separate from TX/R. TX/R can be sited away from the (main) operating position, convenient for the AMU.

5.1a Standard one-piece unit transceiver with fixed control panel. TX/R must be sited at the operating position.

Fig 5.1 Marine SSB Installation

HF services provide two-way communication, from ship-to-shore and inter-ship, on the high seas, and from the waters of countries other than your own. HF effectively provides world-wide communications through a number of long-range coast stations. HF is not suitable for shorter range contacts. Using sky-wave, you skip over stations near to you to reach those further away.

Medium-Range '2MHz R/T Band'

The Medium Range (1.6—4MHz) radiotelephone band (sometimes referred to as the '2 Megs Band') has been around for many years, much longer than VHF, but can now be considered as an extension to VHF. Many operators providing an MF radiotelephone service will also provide a VHF service and you should attempt contact on VHF, when in range, before resorting to MF.

Distress and Calling Channel — 2182kHz

The 2182kHz international distress and calling channel is the key to the medium-range R/T service. If you need to call any station and there are no arrangements for calling direct on a working channel — then 2182kHz is the channel to use. 2182kHz performs the same function on MF, as CH16 does on VHF.

Silence Periods

All vessels equipped with MF radiotelephone are obliged to observe silence periods twice each hour, on the hour/half-hour, for three minutes at a time — on 2182kHz.

The reason for silence periods is to allow distress messages to be heard even in busy areas and when low power equipment, like that used in survival craft, is being used.

It is an offence to transmit on 2182kHz during the silence periods, *except* in a distress situation. This is different to Channel 16 VHF, where mariners in busy areas can be confronted with a constant babble round the clock.

If you fit a Marine SSB set, be sure to mark your clock with the silence periods and check the accuracy of your clock regularly.

Distress Calls VHF/MF/HF

If you ever need to put out a distress message on MF, you would use 2182kHz. The procedure is exactly the same as on VHF Channel 16 — namely (i) select the frequency (2182kHz) (ii) select *high power* on the set (iii) use the procedure shown at the end of Chapter 2 to transmit your MAYDAY call and message.

If you have both VHF and MF onboard and your VHF CH16 distress call has not been acknowledged by a Coast Station, you should broadcast also on 2182kHz. Do this even although your Channel 16 call has been acknowledged by another vessel — unless your rescue is imminent. If your VHF distress message has been acknowledged by a Coast Station, Coastguard, Marine Operator or Port Authority, you can normally rely on them to arrange for further broadcasts on 2182kHz if necessary.

Control of Distress Communications

The station in distress is the control station for distress communications on VHF CH16, 2182kHz or the HF frequency used — as long as they are able to exercise that control.

If a Coastguard or Coast Radio Station has acknowledged the distress, the casualty can ask them to assume control of communications.

If the casualty loses the ability to communicate, then control reverts automatically to the Coast Station involved, on appropriate frequencies.

Where VHF is being used for on-scene communications and the casualty is out of range of a shore station, another vessel might assume 'on-scene' radio control.

In some cases, especially major disasters, the Coastguard may appoint an 'on-scene commander' to control the rescue — including control of casualty communications.

If you have tried VHF CH16 and 2182kHz to no avail, with no acknowledgement from ship or shore, then you need to try something different.

Even if you have activated your EPIRB, you still want to establish two-way contact, up until the time you are taken off or until you have to abandon. That is where the HF distress and calling frequencies come in.

The US Coastguard and the Canadian Coastguard maintain a radio watch on MF and HF frequencies, in addition to the CH16 VHF watch, from a number of stations providing coverage of high seas, coastal, Great Lakes and navigable inland waterways.

Other countries provide an HF distress monitoring service for DSC calls only, as part of the GMDSS (Chapter 9). HF stations used to listen on common calling channels for ships requiring a radiotelephone channel. That is no longer the case. When you are in trouble, far from shore, and have had no response on the designated distress frequencies, you need to know which station is listening where and who you can reach.

If you have built up a log of stations you can hear from different sea-areas, at different times of day, you will be surprised at how easy it is to pick the right frequency — one which gets you the response you want, ie

'Mayday Foundering Maid *this is Halifax Radio — Received Mayday — Standby'*

Ending Distress Communications

The rescue launch has picked up all five people from the foundered yacht. They are in various stages of shock, but otherwise unharmed. The launch has told the Coastguard on VHF that they are two miles from harbour and will be entering shortly.

The Coastguard will have assumed control of casualty communications and is therefore responsible for cancelling the Mayday. Once assured that all people are accounted for, he will broadcast on all channels used for distress traffic.

'Mayday all ships this is Halifax Radio — Mayday Foundering Maid *— All five persons rescued — vessel sunk — distress traffic ended — Seelonce Feenee'*

Medium-range Radiotelephone Service

In the USA they are 'Public Regional Coast Stations'. In Europe and elsewhere, they are 'MF Coast Radio Stations' and they provide a radiotelephone service in the 1.6—4MHz band.

Some stations provide the service on simplex channels (ship and coast stations both using the same frequency) — others on duplex.

Some stations always use the same frequency (ship/coast station) 'pairs'. Others will chop and change, mix and match, from the frequencies allocated to their station.

As most modern Marine SSB sets are fully synthesized and can produce any frequency combination (transmit and receive pair) required, the modern mariner should have no trouble in working into any shore station.

Some stations publish frequency pairs for direct calling from ships of the home nationality, but maintain a watch on the international calling channel 'for foreign vessels'. This is because many older ships are fitted with equipment which will not tune to all channels and who therefore need to call on 2182kHz and be sent to a channel where they can be worked.

If you are a 'foreign' vessel (eg a US vessel off the UK coast; a UK vessel approaching the USA) and you have a fully synthesized set, you will not be admonished for calling direct on a working channel, where direct calling is the norm for vessels of the 'home' nationality.

The frequencies in use for stations in the MF R/T service can be found in ALRS Vol 1, along with any special calling arrangements.

Long-range (HF) Bands

Long-range (HF) Radiotelephone Service

The same SSB set that gives you access to the medium-range R/T service will also provide access to the long-range service, (the 'High Seas' service in the USA), operating in the high-frequency bands.

You have a tremendous choice of stations in many different countries, with the long-range service.

In terms of call charges and convenience, it is often better to contact a station in the country to which you want to make your telephone call. So, if you are calling someone in Canada, you could choose to call Halifax or Vancouver. For a call to the UK you would go directly to Portishead Radio.

In the USA, there are a number of stations, operated by different companies, ready to give you service

Regardless of which country you might want to speak to, you might choose to call the HF station in your own home country and get them to connect you via International Direct Dialling. Not a problem for most developed countries and it could make your accounting easier.

A small sample of HF Radiotelephone Stations is listed in Appendix E.

```
┌─────────────────────────────────────────────────────────────────────────┐
```

Direct Calling on the HF Radiotelephone Bands

Many long-range R/T stations — including Portishead Radio (UK); AT&T 'High Seas Service' Stations KMI, WOO and WOM (USA); and some of Australia's HF RT Stations — maintain a watch on one or more working channels in each band, for ships to call direct on those channels. As there may only be one operator to monitor a number of channels you need to indicate which channel you are calling on, eg

'WOM Channel Four One Two, WOM Channel Four One Two, this is British yacht *Albert Ross*, *Albert Ross*, call-sign Golf Uniform Lima Lima, off Daytona Beach, over'

or

'Portishead Radio, Portishead Radio, Portishead Radio this is *American Liberty*, *American Liberty*, *American Liberty* — WYZ666 calling on Channel GKU46, over'

Note the different styles preferred by AT&T and by Portishead Radio. Portishead Radio use the international calling style, whilst AT&T High Seas stations have adopted their own procedure. But both methods do contain the same basic information. (Specific station details are included with the frequency table in Appendix E.)

```
└─────────────────────────────────────────────────────────────────────────┘
```

Inter-ship

There are inter-ship channels on both MF and HF (Appendix F). These allow you to communicate over much longer ranges than on VHF, although you will probably have to make arrangements with friends as to when you will both be listening and on which frequencies. Worth considering though, as you want to get the most from your SSB installation, as well as putting it to the test at regular intervals and to confirm the ranges you can achieve on various frequencies, rather than wait until you *really* need it.

But do not forget that marine radio frequencies, other than those public correspondence telephone channels, are meant for 'ships business'. This will allow you to exchange positions, passage details, weather conditions and rendezvous dates and times — but do not dwell on social calls about the party you had when you were last together, if you value your licence!

Selecting Your Frequency

I mentioned earlier that you should attempt VHF contact when in range of a Public VHF Station before moving to MF. It is also normally advisable to use MF rather than HF, if only for the price difference which exists in many areas for MF/HF RT calls.

When it comes to HF, you might have a number of stations which appear equally suitable, each of which operate on the whole range of HF frequency bands. Even if you are sure of the station you want to work — how do you choose which *band* to call on when the station offers everything from 4MHz to 25MHz?

Firstly, consider some general guide-lines.

During the hours of darkness you will only get service on the lower HF bands (4, 6 and/or 8MHz).

As we approach noon and the first couple of hours of the afternoon, the ionosphere is coming up to maximum ionisation. This usually means that the lower HF bands (4MHz, 6MHz and 8MHz) have been progressively absorbed by the D-layer (see Chapter 2). You ought to be using 12MHz and upwards if you want to achieve long-range communications.

But how do you choose between 12MHz, 16MHz, 18MHz, 22MHz and, in the extreme, 25MHz?

Swing through the bands to find which channels are open and appear to be giving out a good solid signal from your chosen coast station. You should then pick from the two highest frequency bands which are open and are giving a solid (ie — without occasional fading) signal.

Whether you pick the highest or the second highest HF band which you can hear, will depend on the time of day *at the position mid-point between you and the station you want to work*. Easy to decide if the station is North/South of you, but needs a little thought if East/West.

Use the *highest* frequency which is open to you if the mid-point position has not yet reached maximum ionisation (Chapter 2). As ionisation increases, the optimum working frequency will get higher, so choosing a lower frequency could leave you with a gradually weakening signal.

If you are reaching or passing the point of maximum ionisation, (early afternoon) you might choose the *second highest* frequency. This would normally give you good solid copy for a reasonable period of time, before the ionosphere again becomes too weak to refract that frequency.

Once your radio path has passed the point of maximum ionisation the Optimum Working Frequency (OWF) will be coming down the band and you need to be in front of that change if you want to get the best path.

Do not forget — the best HF station to work is not necessarily the nearest one to your position. There could be a station much further away, with a better propagation path.

Broadcast Services — Weather and Navigation Warnings

One of the best ways of understanding HF propagation is to monitor the various stations which provide a weather and navigation warning service.

You can also monitor traffic lists, broadcast at regular intervals from stations providing a radiotelephone service, to see which stations are on a good path.

Since you are likely to be listening into broadcast stations to gain weather and navigation information, a sample 'watchkeeping schedule' is included in Appendix N2 to help you to keep track of the stations to be monitored. You can also note the quality of the signals received, so you will have a good idea who you can work when you have to. If you contact the service providers direct, they will provide details of any other stations they operate, broadcast schedules and link-call charges etc.

General Calling Procedure in the MF Radiotelephone Band
(Where the Coast Station being called does not use 'direct calling' on working channels.)

1 Check the table of working frequencies (transmit and receive) for the station you are about to call — and note which Channel(s) you can use.

2 Listen on 2182kHz until you are satisfied that there is no distress working (eg — you hear normal calling/answering, and/or broadcast announcements).

3 Check your clock — to make sure you do not transmit during the 'silence period'.

4 Select 2182kHz on your transceiver, and high/low power
(low if close to the Coast Station, high if required because of greater range).

5 Call the station you require, using the standard international calling procedure, eg

'*Nassau Radio this is* Albert Ross, Albert Ross — *Golf Uniform Lima Lima — one radiotelephone call — 2126 — over*'

and expect to hear

'Albert Ross *(this is) Nassau Radio — 2126, listen 2522 and standby*'.

'*Standby*' means that Nassau will call you on the agreed frequency when ready. Had he said '*turn number one*' — then you would be called in after the vessel already working that channel (and so on for turn two, three etc). When called in on the working channel, be prepared to state your details (vessel name and call-sign, accounting authority etc) and the telephone number you require. You will then be connected on your call.

Entertainment and News Broadcasts

Some people like to get completely away from the politics of home — others like to know what is going on all the time. Or they like to know what the media *says* is going on.

If you want to listen to those newscasts from home, or even to hear what is being said from the country you are about to visit, then you can listen to the long-range broadcasting services from the country of your choice.

Marine SSB sets are restricted to particular frequencies for transmission, but the receiver can usually scan through the complete range from the lower frequency (long-wave) domestic broadcast frequencies to the top of the high-frequency, world-wide bands. This allows you to tune in to all sorts of additional non-maritime stations.

A sample of Long-range broadcast (news and entertainment) stations for a number of countries is listed in Appendix K.

Frequency Standard and Time Signals

A number of stations broadcast frequency standard and time signals on the HF bands.

Notable amongst them are the broadcasts of WWV/Fort Collins, Colorado — which covers the waters of North-West Atlantic, the Caribbean Sea, the Gulf of Mexico and the North-West Pacific.

WWV broadcasts brief storm warnings and wind speed, in addition to time signals.

Appendix G covers stations operating a frequency standard and time signal service.

Installing Marine SSB

The three main things to consider when installing a Marine SSB set are (i) where do you want to operate the set from? (ii) where is the best place to site your antenna? and (iii) what will you use as the antenna 'ground', or counterpoise?

Operating Position

Unlike VHF, you will not normally be using your SSB set to talk or listen to harbour authorities or other vessels when navigating in confined waters. The need to operate your SSB installation from the conning position is therefore normally limited to emergency communications on the high seas — something which you hope will be a rarity (but which needs to be considered).

For the vast majority of time your SSB set will be used to receive weather and navigational information, as a broadcast receiver, or to make radiotelephone/radiotelex calls — and the most appropriate place to carry out these functions is below-decks. You would want, therefore, to control your SSB set from, or close to your navigation station.

There is a lot to be said for an additional extension loudspeaker and/or a second control unit at the conning position, if it is not too exposed.

Antenna Siting

If you have a mast or masts, you can rig a wire antenna — this usually provides the best (and least expensive) arrangement for sail boats.

This can be done by substituting an insulated backstay for the existing backstay, which is practical and convenient, but seen by some as introducing an extra weak point into the rigging. If you are a mainly fair-weather sailor then this should not be a problem if the job is properly done. If you are inclined to sail in strong winds and rough weather, you might want to consider an alternative antenna rig — like a whip or a wire which does not actually support the mast, or which is not the sole support.

a Sloping Wire (Insulated backstay)

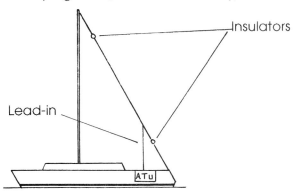

Insulators

Lead-in

b Sloping Wire (Insulated stay)

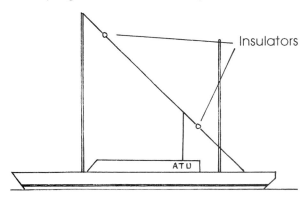

Insulators

Sloping wire radiation patterns are not as efficient in every direction as a vertical wire. The signals will enter the ionosphere at different angles, depending on direction, leading to shorter range and/or reduced signal strength at the receive antenna.

The top insulator should be at least 2ft(60cm) from the masthead. The bottom insulator should be high enough to prevent accidental touching of the radiating portion of the sloping wire. The lead-in should be insulated wire. (But do not forget — the lead-in will also radiate radio waves and can still burn.)

Fig 5.2 Marine SSB Antenna Arrangements for sailboats. (Wire antenna is also suitable for short-wave ham rigs and broadcast receivers.)

c 'T-shape' Arrangement

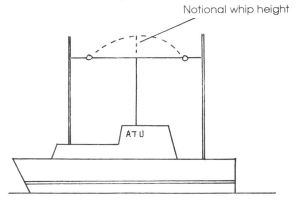

Notional whip height

The capacitance caused by the horizontal top wire has the effect of heightening the vertical wire, making a more efficient radiator than a whip of the same height.

d 'Inverted -L' Arrangement

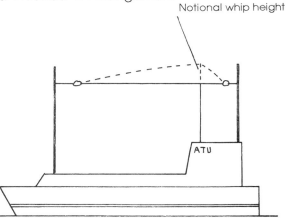

Notional whip height

As with the 'T-shape', the vertical wire is the main radiator and is artificially heightened by the horizontal top wire.

For all wire antennas it is a good idea to use insulated wire where people can touch. A painted red circle of 18 to 24 inches radius around the foot of the wire, where it passes through the cabin roof can also be effective in alerting crew to the potential hazard.

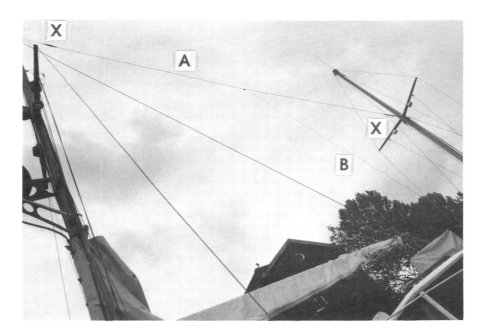

The horizontal wire 'A' has the effect of artificially lengthening the vertical wire 'B', which is the main radiator in this arrangement. Note the insulators at 'X'.

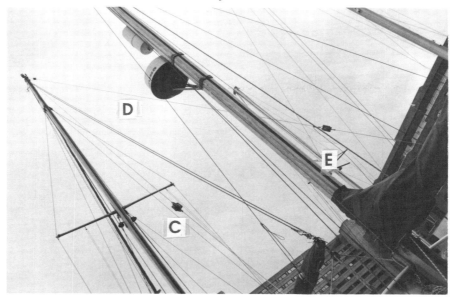

Note the sloping wires at 'C' and 'D' and the lead-in at 'E'.

A wire antenna can be rigged up in a number of ways (See Fig 5.2). The main thing to remember is to rig it clear of sails, booms or anything else which could be fouled. And it must be properly insulated.

The upper insulator of a wire antenna should be at least 2ft from the masthead. The lower insulator should be at a height which will prevent anyone accidentally touching the uncovered radiating portion of the wire (radio frequency burns can be extremely painful).

The lead-in from antenna to ATU should be insulated and as short as possible. The ATU itself, therefore, needs to be as close to the base of the antenna as is practicable. There are ATUs which can be fitted above-decks and others which must be protected from the elements. It may be impractical on a sail boat, using a wire antenna, to fit an external ATU.

It is also desirable to fit the transceiver close to the power supply (battery) and to keep the distance between the transceiver and ATU as short as possible. This is one reason for choosing a transceiver with a separate control panel, especially on larger boats.

Power Supply

Marine SSB equipment normally operates at a nominal 13.8 volts DC. This is the voltage level of a '12-volt' supply when fully charged.

As your SSB set will draw a significant current when in transmit mode, it can quickly pull your supply voltage down — leading to reduced radio performance.

When using your SSB to transmit (or when using *any* power-hungry electronics) you should run your generator to maintain the supply voltage level and battery charge.

Antenna Ground/Counterpoise

You will remember that on MF/HF rigs we use a quarter-wave antenna (Chapter 2) — but that we really need a half-wave arrangement. Also that the antenna ground, or counterpoise, forms half of the antenna arrangement (The 'Earth Mat' in Fig 5.1).

The antenna ground needs to be a substantial area of metal close to, or in direct contact with, the sea. The electrical 'earth' system onboard your boat is *not* suitable, on its own, as an antenna counterpoise.

The hull of a metal craft is ideal. If your vessel is a metal one, you should connect the ATU directly to the hull, using broad, flat copper strip. This should be attached to a metal stringer with a strong bolt — after cleaning the area of paint or other covering.

Separate the copper strip from the metal of the hull with a stainless steel washer, to reduce the likelihood of electrolysis, and then paint over the connection, or cover it with petroleum jelly to keep it free from moisture.

Wood, fibreglass and cement hulls are not so easy. None of those materials will do as a ground plane, so you need to somehow introduce one onto your boat. This could mean considerable upheaval.

Electrolysis — The Unwanted Battery

When you place any two dissimilar metals against each other, those metals will form a natural electrical cell (a small battery) — and a small current will flow from one 'plate' to the other. The amount of current flowing depends on the metals used and the environment (air, salt-water, fresh-water etc). If the current flow is excessive for either type of metal (aluminium is particularly vulnerable), then the plate will be eroded — just as in a real battery. A 'Noble Metals Table' will detail the maximum current allowable for any two metals, to help you avoid burning holes in your hull!

The counterpoise for a standard 150 watt, 'small ships' transmitter should ideally be an expanse of metal of at least 10sq ft/1sq m, close to or in contact with the water. The trend nowadays is to have a sheet of metal on the outside of the hull (glued, bonded or inset) faired if proud. This arrangement will give you the most consistent performance across the MF/HF bands. The alternative position is inside the hull, below the waterline. The technician installing your set will therefore need time and space to get the mat in place — and this may well involve the temporary removal of furnishings and fittings.

The best counterpoise is the *solid* metal mat. It can, however, be made with fine wire mesh, like chicken wire, or bronze window screen. Whatever you use will need to be bonded to the inside of the hull and any open edges soldered to make the mat complete.

Connect the counterpoise to your ATU with broad (2in or wider) copper strap. The longer the strap, the wider it should be.

Although not in direct contact with the water, the radio frequency signal will have little trouble in bridging the gap between an earth mat and the sea — this is because electromagnetic waves at radio frequency would see the mat/hull/sea as a huge capacitor — capable of stopping Direct Current (DC) electricity, but virtually a short circuit to Alternating Current (AC) at radio frequencies.

Mesh type mats will have different electrical properties at different frequencies, so performance will vary according to the frequency band in use.

Lightning Protection ?

On the subject of Direct Current — although your mat/hull/sea sandwich would block 12 volt or 24 volt electricity, it will not stop lightning from identifying your mast, with or without a radio antenna, as a shortcut to 'earth'. So you need to take one last precaution.

Your earth mat/ATU arrangement should also be connected to your electrical earth — using 2in copper strip. All large areas of metal should also be connected together in this way — including the engine block, generators, gearbox etc (something which happens naturally with metal hulled boats).

The whole should be connected to a DC earthing block, or similar arrangement, which *must* have a direct connection to the sea. Any radio equipment in the path taken by lightning will be destroyed. This could be the time when you find out that only your EPIRB and hand-held VHF still work.

Whip Antenna

The above arrangement represents the complete MF/HF radio installation for sailing craft. For those without a mast, the configuration is the same except for the antenna itself. Without a mast to support the top end you need to substitute a vertical whip antenna for the 'long wire'.

A whip antenna *can* be mounted on a mast — whether a single post or a tripod type — or it can be mounted on a tower, on top of/attached to the side of the wheelhouse or on the stern rail.

The HF whip — at 17ft to 35ft long — will normally need a support in addition to the single base bracket or it will come down too easily. It is not a good idea to put a whip antenna where it can be used as a grab handle or other form of support.

Radio Frequency/Service Information

Like VHF services, MF/HF stations and frequencies are listed in the ALRS, Radio Aids to Navigation, Notices to Mariners etc. Nautical Almanacs tend to concentrate on short-range services rather than MF/HF.

Summary

Installing an SSB transceiver requires much more thought and effort than a VHF; but a good installation will provide you with long-range access to weather and navigational information, the public correspondence radiotelephone service and friends on other craft at great distances.

It will give you the facility to exchange distress messages at medium and long-range, or to gain advice in a medical or other emergency. You can talk directly to the Coastguard in the USA and Canada if you require their advice or support on the high seas.

If you want to tune into world-wide news and entertainment broadcasts, you can. Your SSB receiver will tune through the whole MF/HF band and will be able to receive the double side-band, (AM) broadcasts as well as Marine SSB services.

So much for the basic Marine SSB without add-ons. The next chapter explains the wealth of additional facilities available on the MF/HF bands and what you need to add to your set to take advantage of them. It also explains the (terrestrial radio) alternatives to the Marine SSB transceiver for long-range reception and/or transmission of messages of all types.

Chapter 6

Marine SSB Add-ons and Alternatives

Introduction
Your Marine SSB transceiver will allow you to monitor weather forecasts etc, when broadcast by voice; to make two-way inter-ship calls and R/T calls to people ashore; and to listen-in to news and entertainment stations. There are also a number of other maritime services available across the MF/HF bands that you might want to use, but which are not accessible with the basic set.

Services like NAVTEX, Weatherfax and Narrow Band Direct Printing (NBDP) Radio-telex — also known as 'SITOR' — all of which you can take advantage of by adding to your single side-band transceiver or, for some, by installing a stand-alone unit instead of, or in addition to, your SSB radio.

NAVTEX, Weatherfax and SITOR broadcasts of navigational and weather information can be monitored by using a dedicated marine device with a built-in receiver and its own independent antenna system, or by using an inexpensive 'communication receiver', or a 'world-band radio' and a multi-system decoder.

Selcall (SSFC) and Digital Selective Calling (DSC) units can be added to your Marine SSB, as can an automatic R/T system (like BT's 'Autolink' or AT&T's 'High Seas Direct', which form part of the 'Global Maritime Radiotelephone Service Group' for provision of automatic radiotelephone services).

But you might choose not to have a Marine SSB set at all. You may feel that you have made adequate arrangements for distress alerting and for passing messages to and from shore by other means — but your interest in radio and what it can do both socially and technically might lead you to fit an amateur (ham) radio set.

This chapter explains the implications of making that choice, what you gain and what you lose, as well as explaining why you cannot (legally) have one set to cover both marine and ham radio.

NAVTEX

NAVTEX is an international broadcast facility for transmitting weather, navigational, safety and SAR information, on a standard world-wide frequency, using English as the common language. Forming part of the GMDSS (Chapter 9) communications infrastructure, NAVTEX equipment is designed to automatically accept some types of information whilst the user can choose to accept or reject other types.

Message Types

The categories of NAVTEX message and their designator letter are:

Designator	Category
A	Navigation Warnings
B	Strong Wind (Gales, Storm, Hurricane) Warning
C	Ice Reports
D	Distress Alerts
E	Weather Forecasts (other than B)
F	Pilot Messages
G	Decca Messages
H	LORAN Messages
I	OMEGA Messages
J	SATNAV (GPS etc) Messages
K	Messages about other radio navigation services
L	Additional Nav Warning Numbering Series (overflow for Category A)
Z	'No Messages on Hand'

(not all categories will be used by all stations)

Messages are numbered sequentially by each station, under each category — eg, HO2 would be the second LORAN message from a particular station. When a station reaches 99 in a particular series then numbering restarts at O1. By numbering messages in this way, the receiver can recognise a message which has been received on a previous broadcast and avoid presenting you with a duplicate message. By preventing duplication and by allowing you to programme your receiver to reject any category of message you do not wish to receive, you avoid having reams of useless paper or seeing your system memory-bank being filled with unwanted traffic.

Category A, B and D messages *cannot* be 'programmed out' however. If your set is switched on and you are in range of a NAVTEX station when it broadcasts a message in any of those three categories, then your receiver will recognise it and capture it.

The original standard of NAVTEX equipment, and all equipment designed for compulsory-fitted vessels, must have a paper print-out. It will also print category A, B and D messages on receipt.

This could be a nuisance on a small boat where you are in a secure berth, at night and asleep, but have left your NAVTEX receiver on to catch the 0600 forecast. Fortunately, there is now a range of NAVTEX receivers which the voluntary-fitted owner can take advantage of, including some which will receive and store messages, with 'instant print' suppressed until you ask for it. By using one of those sets you can sleep soundly throughout the night and still receive your up-to-date weather at breakfast time.

NAVTEX Stations 'Timeshare' Arrangement

The NAVTEX frequency of 518kHz, being common to all NAVTEX stations, is divided into timeslots within each of the sixteen world-wide Navareas.

Stations within any one area receive a station designator (A to Z), allocated in such a way that vessels working near the limits on one area, having selected a particular station on which to maintain watch, should not receive messages from the adjacent area — eg: the most southerly station in Area I is Brest-Le Conquet (France) with the designator 'F'. The most northerly station in Area II is Corsen, with the designator A, whilst 'F' in Area II is allocated to Horta, in the Azores.

NAVTEX stations and their broadcast schedules are listed in Appendix J.

NAVTEX equipment is relatively inexpensive and is easy to install. All it normally requires is a suitable (12volt) power supply and an antenna. The simplest sloping wire antenna will serve well as a NAVTEX antenna or you can use a small (active or passive) whip or a 'port' on a shared active antenna.(See Chapter 3, Fig 3.2 and Chapter 5, Fig 5.2)

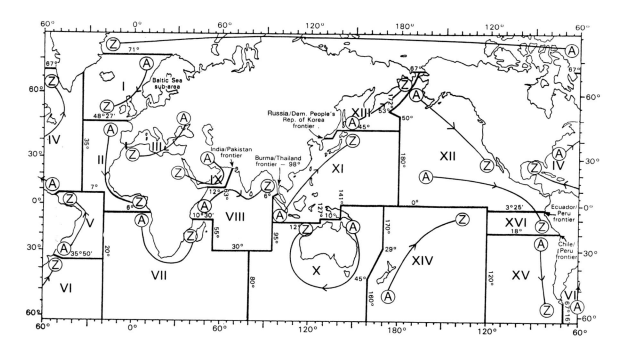

Fig 6.1 NAVAREAS of the world-wide navigational warnings service showing the basic scheme for allocation of NAVTEX transmitter identification characters by IMO.

Weatherfax

There are a large number of stations around the world which transmit weather 'pictures' in the form of a facsimile message, the system being known as 'Weatherfax'.

Like NAVTEX, you can install a dedicated piece of equipment (with its own antenna and printer) to receive Weatherfax broadcasts or you can add a decoder to your Marine SSB set, using the SSB receiver and antenna system as your main signal source. The decoder will need a separate printer to produce the weather picture on paper.

Unlike NAVTEX, each Weatherfax station has its own radio frequency. You therefore need a tuneable receiver and a prepared schedule if you want to be sure to pick-up all the Weatherfax information you require.

When considering a Weatherfax machine, and/or a printer for a Weatherfax decoder, find out the type (and cost) of the paper needed for the print function. Thermal image paper is considerably more expensive than plain paper and what looks like a good price advantage when buying the equipment can quickly be overturned by ongoing costs. If you want a dedicated machine, however — one with its own integral printer/imaging system, you may have no choice but to accept thermal image paper.

A small selection of Weatherfax stations operating in a maritime environment are listed in Appendix H. For a more comprehensive list, see ALRS Vol 3, or other specialist publications.

SITOR (Radio Telex) Modems

The third SSB add-on mentioned is the radio telex facility, known as 'SITOR' (Simplex Telex Over Radio).

You can receive weather messages and navigation warnings using a SITOR decoder — which may be combined with a NAVTEX and a Weatherfax decoder in a single unit. Many stations also use SITOR as a means of continuously sending their radiotelephone traffic list.

But SITOR also offers a two-way ship-to-shore messaging facility, just like the radiotelephone service but into the international telex network. With a SITOR send/receive facility you cannot only pass messages over the telex network, but you can also send 'Radio Telex Letters' to people or places which do not have a telex machine of their own. The radio station automatically intercepts the RTL, prints it out and mails the letter to the addressee.

One particular advantage of having a SITOR facility is the access provided to information databases. Such databases can include the latest weather forecast for a particular area; call charges (telephone and telex); navigation warnings and — which can be particularly useful — Optimum Transmitting Frequency guides. (The HF OTF guides for a particular station will serve for radiotelephone and radiotelex frequencies).

Communications/World-Band/Broadcast Receivers

The term 'communications receiver' should strictly speaking be reserved for high quality, high specification (and therefore fairly expensive) radio receivers covering, at least, the MF/HF frequency spectrum.

'World-Band' receivers also cover short-wave but vary considerably in quality and specification, and in their ability to tune into a particular signal whilst rejecting others.

The cheapest world-band receivers are to be avoided, being only of limited use to the mariner, but there are some reasonably affordable broadcast receivers, no more expensive than the cheaper Marine VHF sets, which will allow you, for a limited outlay, to listen into the same range of frequencies and facilities as a Marine SSB set. You can then judge for yourself whether or not to invest in the (much more expensive) Marine SSB transceiver (or in a ham rig).

There are two additional advantages in this approach.

Firstly, in countries which require you to pass an examination before you are allowed to use Marine SSB (including Australia, New Zealand, South Africa and the UK), you can become familiar with the MF and the short-wave bands long before you are licensed to fit and operate an SSB transceiver.

Secondly, whichever your native country, you can set your broadcast receiver up at home and do your experimenting *before* installing any equipment onboard. In that way, once you do move the set onto your boat, you are not trying to 'learn radio' at the expense of enjoying your time afloat, nor going to the trouble of fitting a suitable antenna before you are convinced that you really want one.

When choosing a broadcast receiver for this purpose (or indeed, if you want to fit a receive-only system to link to a NAVTEX/Weatherfax/SITOR decoder), you ought to be looking for a set which:

— has a socket for an external antenna (you will not do very well with the telescopic whip attached to the set).

— has an output to drive your decoder (if required) and which provides enough output power to do this; and

— can be adapted to work off the onboard power supply.

Amateur Radio Equipment (1)

It will be worthwhile reading the ads in amateur radio magazines and speaking to some of the ham-radio suppliers before choosing between the full-spec marine kit or a lower cost alternative — especially if you can find a supplier who is familiar with both. Because radio hams are interested in all types of radio communication (and marine and aeronautical systems seem to hold a particular fascination for many), the ham market attracts suppliers and manufacturers who are not bound by convention when seeking a solution, and often they come up with a product suitable for the job but without the expensive and lengthy development costs normally associated with marine radio equipment.

The fact is that there are now a number of low-cost systems available which can read and translate Morse code — presenting you with a plain language print; which can decode NAVTEX and SITOR broadcasts; and which can reproduce a fax from a weatherfax broadcast — in some cases at an equipment/software cost lower than that of a marine unit which will carry out only one function.

Such systems might not be suitable for 'big ships' operation but may well be adequate for the needs of yachtsmen and owners of other leisure/fishing vessels.

Automatic Radiotelephone

One recent addition to the marine radio product portfolio on VHF, MF and HF, is the automatic radiotelephone attachment. There are a number of systems operating on VHF (See Chapter 3) but only one, with the possibility of a second now that DSC is becoming more widespread, on MF/HF.

The first 'global' automatic R/T service is provided by the 'Global Maritime Radiotelephone Service Group' (GMRS), a consortium of organisations including AT&T (USA), Belgacom (Belgium), BT (UK), CPRM Marconi (Portugal), Gibtel (Gibraltar), Iritel (Italy), Telkom SA (South Africa), Telefonica (Spain) and Telecom Malysia — all of whom provide service on either VHF, MF or HF or a combination of frequencies from all three media (See Fig 6.2).

Using the generic name of 'Autolink RT' — carried by BT for its UK service (the first automatic provider on medium frequencies) — the service is also marketed under other names (eg. AT&T's 'High Seas Direct').

Users buy an add-on unit which connects to the onboard VHF and/or Marine SSB set and which provides direct access to the shore telephone network through participating coast radio stations.

Each Autolink unit has a unique identification number built-in, which is registered with the various Autolink providers so they can allow you to have access to their system. On registration, users are allocated a PIN (Personal Identification Number) which is associated with the Autolink Unit ID in the CRS equipment — so providing security against unauthorised use of your Autolink unit. (The PIN facility does not interfere with VHF/SSB operation over non-Autolink systems so people can still use the radio for inter-ship and other services).

There is also a 'multi-PIN' facility, whereby you can ask for more than one PIN to be allocated. This could be used, for example, to differentiate between private and business calls; or for separate accounts for individuals onboard the same vessel, or even for charter companies to differentiate between successive hirers (A total of 99 PIN's can be allocated at any one time!).

Autolink RT can also include a 'scramble' facility for the radio link, as an 'extra'. This has proved to be a great attraction for fishermen in the North Sea, who traditionally used all sorts of verbal codes to pass catch details ashore, on the otherwise open channels. Like all 'extras' a unit with the scramble facility is more expensive than the basic unit, but there must be a few members of the British Royal Family who would have preferred a mobile

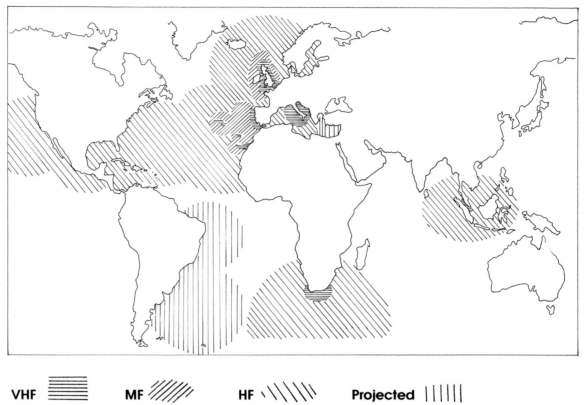

VHF ≡	**MF** /////	**HF** \\\\\\	**Projected** ‖‖‖‖

Map information courtesy of Cimat® SpA

Providers	VHF	MF	HF
South America	—	—	✓
Belgium (Belgacom)	✓	✓	—
Gibraltar (Gibtel)	✓	—	—
Italy (Iritel)	✓	—	—
Malaysia (Telekom Malaysia)	✓	—	✓
Portugal (CPR Marconi)	✓	✓	✓
South Africa (Telekom SA)	✓	—	✓
Spain (Telefonica)	✓	✓	—
United Kingdom (British Telecom)	✓	✓	✓
USA (AT & T)	—	—	✓

Fig 6.2 Autolink RT System — Existing and projected coverage 1994-1995

phone with a scramble facility in the early 90's, rather than the system which they did use, perhaps assuming it was secure?

An Autolink RT unit will also allow you to connect a Group 3 or Group 4 fax machine, or a computer to the radiotelephone channel — and exchange voiceband fax or data calls with the shore. At 1.2/2.4 kbits/second, the system is probably not fast enough for high-traffic ships, but should satisfy the more occasional user who does not want to exchange reams of paper on an hour-by-hour basis.

Finally, some providers of the Autolink service offer access to various additional facilities, using 'short code' access and including weather forecasts and weatherfax databases, news broadcasts, tourist information, port services and direct access to Rescue Co-ordination Centres. Some of the codes in force at the time of going to print are listed in Fig 6.3.

Code/Service	GB	B	Mal	P	Sp
00 High Priority Operator Access	✓	—	—	✓	✓
01 Traffic List	—	—	—	—	✓
11 Routine Operator Access	✓	—	—	—	—
12 Recorded WX Information	✓	—	—	✓	✓
13 Local Land WX Forecast	✓	—	—	✓	—
14 National Land WX Forecast	✓	—	—	✓	—
15 3-day Central North Sea Forecast	✓	—	—	—	—
16 3-day Southern N Sea Forecast	✓	—	—	—	—
17 3-day Northern N Sea Forecast	✓	—	—	—	—
20 Weatherfax Synopsis	✓	—	—	✓	—
21 Weatherfax Prognosis	✓	—	—	✓	—
22 Paging Electronic Mailbox	—	✓	✓	✓	—
23 Data Services VHF	✓	—	—	✓	—
24 Data on MF	✓	—	—	✓	—
25 Data on HF	✓	—	—	✓	—
32 Local Time Announcements	✓	—	✓	✓	✓
34 Tourist Information	—	—	✓	—	—
35 World News	—	—	—	—	✓
36 National News	✓	—	—	—	—

Code/Service	GB	B	Mal	P	Sp
37 'What's in Town'	—	—	✓	—	—
38 Telegram Service	—	—	—	✓	—
40 Telephone Directory	—	—	✓	—	✓
41 Telecommunications Information	—	—	—	✓	—
70 Port Services Information	—	—	✓	✓	—
71 Pilot Services	—	—	✓	✓	—
72 Harbour Master	—	—	✓	✓	—
73 Immigration Department	—	—	✓	✓	—
74 Customs Department	—	—	✓	✓	—
75 Police Department	—	—	✓	✓	✓
76 Port Health Department	—	—	✓	✓	—
77 Shipping Agents Association	—	—	—	✓	—
90 Information on RCC	—	—	✓	—	—
91 Local RCC	—	—	✓	—	—
92 National RCC	—	—	✓	—	✓
93 Local Medico Service	—	—	✓	✓	✓
94 Storm Warning Reports to CRS	—	—	✓	—	—

(Information reproduced courtesy of the Global Maritime Radio Telephone Service Group)

Key:WX= Weather: RCC=Rescue Co-ordination Centre: GB=Great Britain: B =Belgium: Mal=Malaysia: P=Portugal: Sp=Spain

GB, Malaysia, Portugal and Spain also provide a short-code facility for Installation Test Calls by Autolink Distributors.

FIG 6.3 Autolink R/T Short Code Services

Amateur Radio (2)

People can get into quite heated debates at any suggestion of fitting ham radio aboard a yacht or other craft, instead of fitting a Marine SSB rig. And yet, if the money was not available for either and an individual chose to go to sea without any means of long-range radio communication at all, that might be considered to be acceptable!

But the choice *is* there and people are entitled to weigh the 'cons' against the 'pros' and, having considered everything, they are entitled to choose either one against the other.

A road vehicle without a steel safety cage, with no air bag and without an anti-lock braking system may not offer the protection of one with all these features, but we are still free to make the choice and, having made that choice, it is up to us as individuals to conduct ourselves in such a way as to minimise any risk, to ourselves and others, in the way we get from one place to another.

Of course, we cannot predict the impact of other road users — nor can we always predict what will happen at sea. We can take out 'insurance' by joining a roadside recovery club. We can fit a satellite EPIRB and make sure it is registered with the shore authorities. (See Chapter 8). We can make arrangements for others to alert the Coastguard if we do not arrive at our planned destination on time. We can carry a full complement of flares, life-jackets and survival suits. We can fit Marine SSB for contact on 2182kHz and on the HF distress channels in that ultimate emergency — or we can choose not to.

We can gamble on the possibility of never needing to call on the emergency services and choose to spend our money on a ham rig instead. When talking about ham radio in this context, I am referring to short-wave radio-equipment which covers the MF and HF spectrum from around 1.5—30MHz (and down even lower than 1.5MHz on the receiver side). There are other types of ham rig — including VHF/UHF and satellite systems — but they are not real contenders against any marine radio system.

There are two main reasons why Marine SSB and short-wave ham transceivers are in contention and the first one is the cost of installing either rig. If Marine SSB and short-wave ham rigs cost the same as a Marine VHF set there would be no problem. Most people would fit the Marine SSB first and foremost, and would only install a ham rig if they were really interested in radio as a hobby. But they do not cost the same as a Marine VHF. You can buy a dozen VHF sets for the price of one Marine SSB/short-wave ham rig. At those prices, people will normally choose one or the other. And that is where the second point of contention comes in.

The problem is, that a ham rig will let you do many of the things you expect from a Marine SSB set. In particular, it will let you do the things most people do every day at sea. You can monitor the weather forecasts and navigational warnings. You can connect your laptop computer and copy NAVTEX, SITOR and Weatherfax broadcasts. You can listen to Radio France Internationale, Radio Canada International or the Voice of the Resistance of the Black Cockerel.

With a ham rig, you can also talk to other amateur radio enthusiasts the world over — which you cannot do on the air with a Marine SSB set.

But a ham rig will not give you two-way access to the marine radiotelephone service, nor will it (legally) give you talking access on the 2182kHz distress frequency, nor the HF distress frequencies.

This is the real choice — two-way communication within the Amateur Service, or two-way communication in the Maritime Mobile Radio service. And, assuming you will fit one or the other, your are really choosing which facilities you are prepared to do *without* when you opt for one rig against the other.

To do that, you need to have some idea as to what the Amateur Service has to offer, and I would again suggest getting hold of an inexpensive short-wave receiver and start listening to what is going on in both services — maritime and amateur. But don't forget — your choice is between the *two-way* communications of the two services — either type of rig will give you access to all the broadcast services for the maritime, news and entertainment providers.

Also, to be fair to both services, you are also choosing between a functional service (the Maritime Radio Service) and the 'family' of radio enthusiasts (the amateur radio service). Both have something unique to offer.

The frequency tables in Appendix E and F will give you an idea of what to expect from the maritime service. The amateur radio medium and short-wave allocations are listed in Appendix L. Some of the things you might want to consider in the Amateur Service are listed below

DX-ing
Many individuals are on the air with the view to working as many 'DX' (distant) stations as possible — or of clocking up a contact in a country which has few radio hams or is rarely heard on the air for some other reason. By listening to the DX contacts and comparing to call-signs heard with those listed in Appendix B, you will have some idea of the reach attainable by a rig at your own location.

Short-wave Listening
If you decide to experiment with a short-wave receiver, to find out what you can hear and from where, you should not depend solely on the frequencies appended to this book. Buy a copy of the 'World Radio TV Handbook' (WRTH) and use that as your guide to broadcast services in other countries. The WRTH also lists 'DX Clubs' in a large number of countries — there will be one near you, where you can get advice on SWL'ing.

It will not take very long for you to decide whether you want to know more about SWL and amateur radio and all you will have to put out is a bit of time and not too much money. At least you will then base your next decision on more positive information than a brief conversation in the bar at the marina!

Ham Nets
Something you may want to listen into are the regular 'Nets', or Networks, where like-minded hams spend some time on the air together at a regular time and on a specified frequency within the ham bands to exchange information. Amongst those nets are some for

Maritime Mobile operators (which you are considering becoming?), who get together on the air with the help of a few dedicated net controllers in different parts of the world. Some Maritime Mobile nets are listed in Appendix M, to give you a flavour of what is on offer.

Mobile Operation
Operating a Maritime Mobile (ham) Station is normally allowed within the boundaries of your own country and in international waters. When using your kit in a maritime mobile set-up, you should use your normal call-sign, followed by the words 'Maritime Mobile' (voice) or /MM (Morse/TOR working).

If you sail into an international Region (1, 2 or 3) other than your own, there will be differences in the frequencies you can use within the amateur bands. You need to be familiar with those differences before you arrive in the other region.

Using Your Amateur Radio Abroad
When you sail from international waters and into the waters of a country other than your own, you may be prevented from using your ham rig for transmitting. On the other hand, many countries allow the temporary operation of amateur kit in their own country, sometimes by reciprocal arrangement (ie, you can operate in their country and their own amateurs can operate in yours).

You can normally only do this with the prior permission of the other country however, and it would be prudent to write and ask for this permission a few months before you require to operate 'over there'. You will be allocated a call-sign, which may be your national call-sign with the addition of the prefix/suffix of the country being visited, or may be a fresh call-sign of that country.

You will also need a copy of the licensing regulations of the country being visited, so don't forget to ask for those regulations when applying for a licence.

Whilst in international waters you are subject to the radio regulations of your own country (that is, the country of registration of your vessel). When in foreign waters you are subject to both sets of regulations, and so must apply the more stringent interpretation where individual rules differ (eg, with Band Plan allocations).

Third Party Traffic and 'Phonepatch'
Most countries do not normally allow amateur operators to pass 'third party' traffic, nor to use the frequencies of the Amateur Service for passing business messages.

That means that the two individuals at either end of the radio link — the two qualified amateurs who are operating the radio equipment — should not pass on messages from other people at either location. For example, if you are the licensed operator on a vessel and you are in communication with another licensed operator ashore who happens to be a relative of a fellow crewman, you should not pass on the message that 'your Dad sends his love'. That would constitute third party traffic.

On the other hand, the USA allows amateurs a particularly useful facility called 'phonepatch'. Using phonepatch equipment, a shore-based ham can connect the mobile ham into the telephone network and onto a non-ham relative or friend elsewhere in the country.

This is not considered the same as third party traffic, but it should be limited to brief social calls — except in an emergency or potential emergency situation and when no more appropriate method of commercial communication (eg VHF R/T) is available.

If that all seems a bit much, or indeed if it is still not enough, the best thing to do next is to find out more from your national licensing authority and from a national or local amateur radio association. The addresses of those for Australia, Canada, Ireland, New Zealand, South Africa, the UK and the USA are listed in Appendix Q. If the one you require is not listed, get hold of a copy of a 'Ham Radio' magazine from your local newspaper vendor, the address you require will probably be listed there.

Another step which you might consider is to put some time into studying for an Amateur Operator ticket. By the time the qualification has been gained you should be in an even better position to choose — Marine SSB or ham rig, afloat?

Marine SSB/Ham Radio Combination

Marine SSB sets and short-wave ham rigs cover the same overall frequency spectrum, about 500kHz — 30MHz for the general coverage receiver, and 1.8MHz — 28MHz for the transmitter. Often, the same manufacturer makes both types of rig, using the same basic components for the main internal workings of the equipment.

It is often the case that both types of transmitter are inherently capable of transmitting on all frequencies throughout its overall range, but that circuitry has been added to block the production of all frequencies, except those for which the set was intended.

So, why not produce a combined marine/maritime mobile ham set?

There is no real technical problem why a transmitter which covers Marine MF/HF and Amateur Service short-wave frequencies could not be produced. The reasons why it is not done are regulatory — on both the marine side and the Amateur Service side.

International regulations require that transmitters fitted for use in the amateur radio service must not be capable of transmission on the frequencies of any other service and must be used 'exclusively' for operation in the Amateur Service.

On the maritime radio side, regulations for equipment for use by non-specialist operators (ie, those with any lessor qualifications than that required by the Marine Radio/Electronics Officer) dictate that the controls of the set should not require any significant technical knowledge. That is why Marine SSB rigs are limited to RF and AF Gain; SSB switch, which automatically selects USB (no inadvertent LSB selection); simple frequency selection; and noise limiter. And very little, if anything, else.

So the two requirements, that of the amateur 'experimenter' and those of the lay maritime radio operator, conflict. And, for the present at least, even those people who hold operating qualifications for both services cannot legally use the same transmitter for both.

Summary

Marine SSB on its own offers two-way ship-to-shore telephone calls, distress messaging and inter-ship communications. It lets you receive a multitude of shore-to-mobile broadcasts including voice weather and navigation warnings, news and entertainment broadcasts.

You can add an 'Autolink RT' unit to your SSB, to give you automatic access (scrambled if necessary) to the shore telephone networks.

Alternative stand-alone equipment can be used to receive NAVTEX, weatherfax and/or SITOR broadcasts of safety information — or you can add a unit (or units) to your SSB rig to capture these broadcasts (the laptop/notebook computer is set to become the universal answer for many mariners).

You can receive most broadcasts with a relatively inexpensive short-wave 'broadcast' receiver, rather than going to the expense of a full Marine SSB installation — the receiver need not cost any more than a Marine VHF; a fraction of the cost of a SSB rig.

A short-wave broadcast receiver, together with information from amateur radio organisations, will also help you evaluate the relative benefits of Marine SSB and a short-wave ham rig, if you are in the position of buying either one but not both together.

If you do opt for a ham rig against the Marine SSB, you will not be able to make two-way radiotelephone calls over the MF/HF Public Correspondence channels of the Maritime Mobile Service, nor will you be able to send out a distress call on 2182kHz or the HF distress frequencies should the need arise.

On the other hand — and especially if you consider the need to make PC telephone calls as being very rare; and if you are prepared to gamble against the need to ever use 2182kHz/HF distress facilities; and, having looked at what the amateur radio service has to offer, have decided that it definitely meets *your* requirement — then you should not feel guilty about opting for a ham rig against the Marine SSB. Especially if you have taken other appropriate precautions for distress alerting and conduct yourself in an otherwise safe and seaman like manner. (But don't blame me if your gamble does not come off!)

Ham radio will only be a good choice if you are going to use it often. It is only by making regular use of radio equipment that you will become familiar with its capabilities and what it can do at any time of the day or night. In that way you might be able to use it to summon help if you do lose your gamble or if something happens to your boat which has to be resolved by an HF contact.

There are other alternatives to Marine SSB for long-range communications which come into the category of satellite communications.

A Satcom set will cost more than the Marine SSB and ham rig combined, for the present at least, although that will change for the better in the coming years. The next chapter explains the Maritime Satcom environment and what you can get out of it at all levels.

Chapter 7
The International Maritime Satellite System

Introduction

The 'Inmarsat' system currently provides the only commercial, international world-wide maritime satellite communication service in the world. A unique co-operative of around 70 nations, Inmarsat has established, and now operates, a four-station satellite system which covers the main ocean areas of the world. The system provides service to users at sea, on land and in the air.

The Inmarsat service offers voice, fax, data and telex communications using a number of different mobile terminals. It also provides facilities for distress alerting, and for broadcasting of weather and other maritime safety information.

All the standard telecoms services offered by Inmarsat provide automatic connection. The user does not have to select a frequency or channel as with other maritime services — making the services as easy to use as the office telephone, fax or telex — and they really do provide virtually instantaneous connection, of a high quality.

But there is a price to be paid.

Satcom equipment is more expensive than the terrestrial radio counterpart and call charges are generally higher. The radome-covered dish antenna associated with earlier types of Satcom terminal is not suitable for installation on smaller leisure craft. Dimensions are already reducing however, and the next generation of satellites, due for launch in the mid-1990's, will allow even more compact units to be developed and brought into service.

This chapter describes the various types of Ship Earth Station (SES) — Inmarsat-A and B for 'big ships'; Inmarsat-M for craft down to about 35ft in length; and Inmarsat-C, Enhanced Group Call (EGC) receivers and L-Band EPIRB's, the last three being suitable for any size of craft.

Future service provision likely to result from Inmarsat's 'Project 21' — which seeks to provide global voice communication from a pocket-sized telephone unit by the year 2000, is also covered.

But first, a look at the service infrastructure supporting the Inmarsat system.

Space Stations

A 'Space Station' is a satellite — and there are four in operation at any one time to cover the main ocean regions in the world (Fig 7.1). The regions are designated as Pacific Ocean Region, Atlantic Ocean Region East/West and Indian Ocean Region.

The area designated as the Pacific Ocean Region (POR) is covered by a single satellite, as is the Indian Ocean Region (IOR). As you can see from the diagram however, users in the Eastern Pacific get service from the AOR-W (Atlantic Ocean Region-West) satellite. There is also a necessary overlap between regions, to provide service at the higher latitudes.

The single Atlantic Ocean Region (AOR) satellite of the original three-satellite constellation had to carry so much traffic that a second satellite was introduced. Coverage is now provided separately for the AORW (West) and AORE (East Atlantic), providing

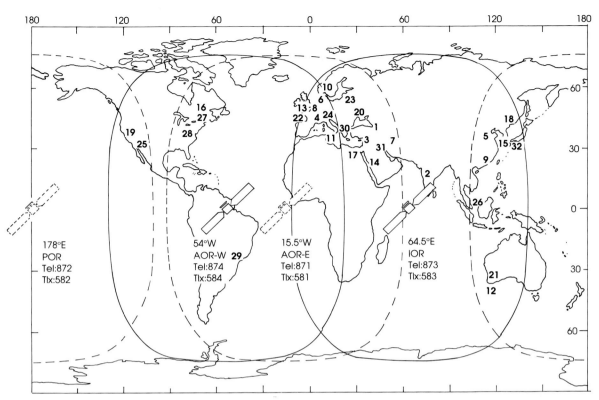

The four Inmarsat satellite ocean regions and the Inmarsat CESs.
(Based on information provided by courtesy of Inmarsat.)

Map labels:

- 178°E — POR — Tel:872 — Tlx:582
- 54°W — AOR-W — Tel:874 — Tlx:584
- 15.5°W — AOR-E — Tel:871 — Tlx:581
- 64.5°E — IOR — Tel:873 — Tlx:583

Satcom Stations and Access Codes for Inmarsat Services (from ship)

Station & Map No	AOR-W M	AOR-W C	AOR-W B	AOR-W A	AOR-E M	AOR-E C	AOR-E B	AOR-E A	IOR M	IOR C	IOR B	IOR A	POR M	POR C	POR B	POR A
1 Anatolia												01				
2 Arvi										306		06				
3 Ata								10/08		310		10/08				
4 Aussaguel	011			011	011			011	011			011				
5 Beijing										311		11/09		211		11/09
6 Blavand						131										
7 Boumehen										14/12						
8 Burum					012	112	012	12/10	012	312	012	12/10				
9 Cape d'Aguilar										118	118			118	118	
10 Elk								04	004	304	004	04				
11 Fucino						105		05								
12 Gnangara																
13 Goonhilly	002	002	002	02	002	102	002	02								
14 Jeddah										315		15/13				
15 Kumsan														207		04
16 Laurentides																
17 Maadi								03								
18 Nakhodka														212		12/10
19 Niles Canyon				13/11												13/11
20 Odessa						107		07		307		07				
21 Perth									222	302	222	02	222	202	222	02
22 Pleumeur Bodou		011		11/09		111		11/09								
23 Psary						16/14				16/14						
24 Raisting						115		15/13								
25 Santa Paula													001	201	001	01
26 Sentosa										210	210		210	210	210	10/08
27 Southbury	001	001	001	01	001	101	001	01								
28 Staten Island								13/11								
29 Tangua						114		14/12								
30 Thermopylae										305		05				
31 Umm-Al-Aish								06								
32 Yamaguchi									003		003	03	003		003	03

Fig 7.1 Inmarsat 4 Ocean-region Coverage

considerable overlap in the area. As each of the Inmarsat-2 satellites has a call capacity equal to 250 simultaneous telephone calls, the overlap substantially boosts channel availability in the region.

Another effect of the re-alignment of the satellites to make four regions was to improve the coverage towards the North and South Poles. Official coverage is stated to be 70° North to 70° South but with only three satellites, the coverage of North-East Canada and the North Cape of Norway was particularly suspect.

The Inmarsat Satellite Constellation

Early diagrams of the Inmarsat satellite constellation show only three birds and three ocean regions — Atlantic, Indian and Pacific.

A second satellite was introduced over the AOR during 1994 to improve the coverage at northern latitudes — particularly for north Norway and north-east Canada.

Introducing the extra satellite required the existing AOR bird to be re-positioned to a point further east than its original position. This operation was carried out by the Inmarsat Operations Control Centre in London, England.

The new (AOR-West) satellite was positioned to the West of the original single AOR bird and now, together, the two (AOR-W and AOR-E) satellites provide considerable dual-coverage across much of the AOR region and much improved peripheral coverage at higher latitudes.

Inmarsat Operations Control Centre

The satellites themselves, their physical positioning and alignment, are controlled by the Inmarsat Operations Control Centre (OCC) in London, England. (See photo). The satellites have been placed in geo-stationary orbit. They are orbiting 22,200 miles (36,000km) above the equator and moving at the speed of the earth's rotation — so that they appear to be stationary above the same point all of the time.

The OCC can operate small rocket thrusters onboard the satellites to maintain their position. The OCC can also bring into service standby satellite units, also permanently in position, in the event of failure of the main bird.

Third Generation Satellites

A new generation of Inmarsat satellites, the third generation, comes into service in 1995/98. These new satellites will utilise 'spot-beam' technology (Fig 7.2) to divide each ocean area up even further, allowing the same frequencies to be used for a number of channels simultaneously. Together with other improvements, this increases the number of voice channels available on a single bird from 250 (2nd generation) to 2,500.

Spot-beams will allow lower power to be used, leading to a further reduction in the physical size/weight of the antenna. Full facility units — offering telephone, fax and data transfer — will therefore be possible for smaller and smaller craft. The increased channel/lower power will also result in a further reduction in call charges, making satellite systems ever more competitive with their terrestrial counterparts.

The new Satellite Control Centre at Inmarsat HQ, London — ready for the next generation of satellites and the new 'Mobile Satellite Service.

Spot-beam coverage using 19 spot-beams and one main beam. The 19 spot-beams share 3 frequency channel groups (Groups A, B and C). By keeping all the 'A's away from each other (also for 'B's and 'C's), the same frequencies can be re-used in one satellite. The main beam carries the signalling channels for whole of the area covered and can also carry some communication channels for the areas between spot-beams, or around the edge of the area.

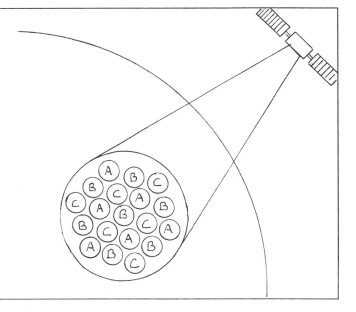

Fig 7.2 Spot-beam Coverage

Service	Short-code Access No	Remarks
Medical Advice	32	Use this code to obtain medical advice. (Some CESs accept this code to make direct connections with local hospitals.)
Technical Assistance	33	Use this code if you are having technical problems with your Inmarsat-M SES. Technical staff at the CES may be able to advise you.
Medical Assistance	38	This code should be used if the condition of a sick or injured person onboard requires urgent evacuation ashore, or requires the service of a doctor onboard your vessel. (This code will ensure that the call is routed to the appropriate medical authority ashore to deal with the situation.)
Maritime Assistance	39	This code should be used in Distress/Urgency situations.

The above is reproduced courtesy of Inmarsat.
(Not all CESs may provide all the services listed.)

Fig 7.3 Inmarsat Short-code Service. 'Inmarsat-M Safety Services'.

Land Earth Stations

The LES, or Coast Earth Stations (CES) as they are called in the maritime world, form that part of the infrastructure which interconnects with the land telecommunications networks on the one hand, and the satellite/earth link on the other.

As well as interconnection with telephone, telex and data communications networks, some CESs also provide direct access to doctors, coastguard or the Coast Earth Station operations staff — where you can obtain assistance in the case of a maritime emergency or when your equipment is malfunctioning. (A small selection of codes for the Inmarsat-M service is shown in Fig 7.3. A, B, C and M providers offer a range of services of this nature — some chargeable and some free.)

Ship Earth Stations

To access the various services, you need to be able to communicate with the satellite. This is achieved by using a Ship Earth Station (SES). The type of SES you choose will determine which services (telephone, telex, packet data network etc) you can access and you will be able to choose which CES you work through, from those serving your ocean area and providing your chosen service.

There are currently six different types of SES, each offering access to particular facilities. The type(s) you choose to fit will depend on (i) what teleComs/dataComs service you want to use; (ii) the size of your craft; and (iii) the depth of your pockets (or, to put it another way, on how much you are prepared to pay for the instant access, ease of use and high quality of service offered by satellite communications).

The available SES types are: Standards A; B; M; and C; EPIRB; and Enhanced Group Call (EGC) receiver. Approximate weights and dimensions are shown in Fig 7.4(a). The facilities they offer are summarised in Fig 7.4(b), and are described below.

Inmarsat-A and B SES

Inmarsat-A was the original 'big ships' unit and continues to offer telephone, fax and telex from the basic unit, as the standard service provision .

Too large for most pleasure craft, Inmarsat-A units have nevertheless been fitted to larger yachts — some ocean racers have used them for prestige events — using some novel methods of lowering the centre of gravity of the antenna (see photo).

This siting of an Inmarsat-A antenna keeps the centre of gravity low — and you will not lose your antenna with the mast! Everyone aboard should be acquainted with the hazards of micro-wave radiation from the antenna. (Photo courtesy Inmarsat)

Inmarsat terminals: size and capabilities

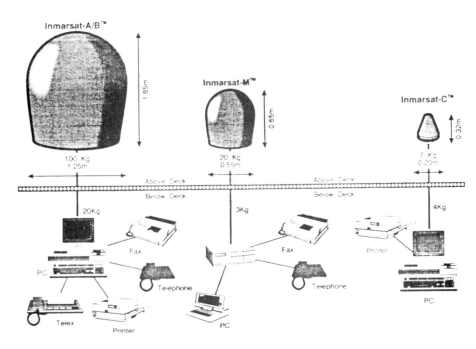

NB: Actual sizes, weights and service configurations vary between models.

Fig 7.4a Inmarsat Terminals (Courtesy of Inmarsat)

Using analogue transmission, Inmarsat-A requires significant bandwidth from the satellite, resulting in the highest level call charges for any maritime satellite service.

A High Speed Data option is now available as an upgrade on some Inmarsat-A terminals. Large data users (eg the offshore oil industry) can now shift considerable amounts of data without tying-up a channel for long periods — to the benefit of all users.

Inmarsat-B is also a 'big ships' unit which, using digital technology, requires considerably less bandwidth than Inmarsat-A — which it will ultimately replace — resulting in a significant lowering in call charges.

Inmarsat-B also offers High Speed Data (from late 1994/1995).

Inmarsat-A and B both offer instantaneous 'Distress Alert' and two-way casualty communications on both voice and telex and are accepted fittings as part of a GMDSS installation (See Chapter 9).

Services	Inmarsat-A	Inmarsat-B (Note 5)	Inmarsat-C	Inmarsat-M (Note 5)
Voice	Yes	Yes	No	Yes
Telex	Yes	Yes	Yes	No
Group 3 fax(rates)	To 9,600 bits per second	9,600 bits per second	No	2,400 bits per second
Data rates (Note 1)	To 9,600 bits per second	To 16,000 bits per second	600 bits per second	2,400 bits per second
X-25 (Dedicated data channel)	Yes	Yes	Yes	Yes
X-400 (Electronic mail box)	Yes	Yes (enhancement)	Yes	Yes (enhancement)
High speed data	56/64 kilobits per second	56/64 kilobits per second	No	No
Full motion 'store & fwd' video	Yes	Yes	No	No
Short Data/ Position Reports	No	No	Yes	No
Group Call (Note 2)	Yes	Yes	Yes	Yes
SafetyNET (Note 3)	Yes, with Inmarsat-C/EGC receiver installed	Yes, with Inmarsat-C/EGC receiver installed	Yes	Yes, with Inmarsat-C/EGC receiver installed
FleetNET (Note 4)	Yes, with Inmarsat-C/EGC receiver installed	Yes, with Inmarsat-C/EGC receiver installed	Yes	Yes, with Inmarsat-C/EGC receiver installed
Distress and Safety				
GMDSS compliant	Yes, if properly installed	Yes, if properly installed	Yes, if properly installed	No
Distress button	Yes	Yes	Yes	Yes

Notes 1 Data Rates: Higher rates may be achieved with data compression techniques.
2 Group Calls: Simultaneous broadcasts to selected groups of users or geographic areas.
3 Services broadcast include distress and safety information, weather and navigational information for fleet management.
4 For fleet management, subscription services like news and other commercial applications.
5 Recently introduced: full range of services not available at time of publication.

Fig 7.4 b Inmarsat Facilities *(Courtesy of Inmarsat)*

Inmarsat-M SES

Inmarsat-M, introduced in November 1992, is the unit which brings satellite telephone and facsimile communications to the leisure craft and fishing vessel owner.

Using a much smaller antenna than Inmarsat-A and B, Inmarsat-M terminals are considered suitable for craft down to 35ft (12m) in length. Ideal for normal telephone and fax exchanges, units also include a 'distress priority' call facility. Inmarsat-M units are not 'GMDSS approved' however, and will not form an essential part of any compulsory fitting on passenger vessels etc. This should not deter voluntary-fitted craft from including Inmarsat-M, especially where additional arrangements for distress alerting have been made.

The introduction of the third generation of Inmarsat satellites promises even smaller units which will be able to be fitted on the smallest of craft.

With equipment prices starting where more expensive Marine SSB units stop, satellite telephone communications using Inmarsat-M is expected to impact the higher end of the leisure market first. Satcom will certainly become much more common-place on larger yachts and boats, commercial fishing vessels and river craft. Especially because SatComs can be used without any real training, passing of exams nor understanding of how the signal gets from one end to the other!

Call charges on Inmarsat-M, like B, a digital system, are significantly lower than Inmarsat-A and are expected to cost about the same as an HF radiotelephone call during the later half of the 90's decade.

Satcom Antenna — Inmarsat-A, B and M SES

Inmarsat-A, B and M units all require a stabilised directional antenna, enclosed in a 'radome' above-decks. The antenna is connected to a gyro compass which helps it to 'track' the satellite in your ocean region. The Inmarsat-M antenna is notably smaller than the A/B unit, which is why it can be fitted on smaller craft.

One manufacturer has gone away from the traditional dome-shaped cover, to produce an antenna unit suitable for fitting onto any flat surface.

The below-deck equipment for Inmarsat-M consists of the Comms unit, about the size of a Video Cassette Recorder, and a telephone/fax machine.

An Inmarsat-M set will draw 6 — 8amps from a DC24V source — about twice that of a Marine SSB set. Like SSB therefore, you should keep your genny running when transmitting.

117

The smaller Inmarsat-M antenna requires less radiation-safe distance than the Inmarsat-A. (About 2 metres is recommended for M, as against 11 metres for A.) Kit can be mast or cabin-top mounted.

Inmarsat-M below deck. Equipment consists of a single unit about the size of a VCR and a telephone handset. Slow-speed fax (2.4kB/s) can also be sent over Inmarsat-M channels.

How to make a distress call using Inmarsat-M SES

1 Lift the telephone handset and listen for the dialling tone
(or switch the handset to the TALK position).

2 Lift the flap over the Distress button and *press and hold down*
the Distress push-button for **at least 6 seconds**.

3 Press the # key to initiate your call.

4 When the Rescue Co-ordination Centre (RCC) Operator answers,
speak clearly and give the following information:
- The IMN* for the telephone channel on your SES
  and the ocean region selected on your SES
  to enable the RCC to call you back.
- Your ship's name and identity.
- Your ship's latitude N/S and longitude E/W.
- Your ship's course and speed.
- The nature of your distress circumstances, for example:
  - Fire/explosion
  - Flooding
  - Collision
  - Grounding
  - Listing
  - Sinking
  - Disabled and adrift
  - Abandoning ship
  - Piracy attack

- The assistance you require.
- Any other information required.

5 Follow the instructions from the RCC Operator
and when requested replace the handset to await further calls.

6 Keep the telephone line clear so that the RCC can
call you back when necessary.

*IMN= Inmarsat Mobile Number

(Reproduced courtesy of Inmarsat)

Power Supplies

Equipment specs will often state that a unit will work from various different sources of supply — eg DC12V, DC24V, AC110/220V. It is usually the case that, whatever the source supply, the equipment operates internally at a low (eg 2 — 12V) DC value.

Where a DC24V supply is available, a DC:DC converter drops your 24V down to the level required. A similar arrangement rectifies AC supplies and produces the required DC level.

Once you have decided to fit various pieces of electronic equipment, especially if radar, radio or satellite transmission equipment is involved, it is better to fit a substantial (DC24V) battery supply. You will also need a generator or other power source to charge/recharge the battery.

The reason for recommending the larger (DC24V) supply is to minimise the current drain when on load eg:

A transmitter which can operate on either 12/24V (direct or via a connecter) and which draws 4 amps of current from a 12 Volt supply, will only draw 2 amps from a 24 Volt supply. A 40 Ampere Hour, 24Volt supply will therefore, sustain your transmitter for around twice as long as a 40AH 12 Volt supply — ie

$$\frac{40AH}{2A} = 20 \text{ hours} \qquad \frac{40AH}{4A} = 10 \text{ hours}$$

This principle applies to all your electrical/electronic equipment — and the more you have, the greater the need for a robust source of supply.

Do not forget — for radio and satellite communications equipment which requires Data Terminal Equipment (DTE — the computer!) and a printer, the quoted current drain for the Comms equipment will not include the DTE/printer power requirement.

Inmarsat-C SES

The smallest two-way messaging unit operating into the Inmarsat system is the Inmarsat-C SES. Inmarsat-C operates a 600 bit/second data communication facility which can send and receive messages from and to subscribers over public data networks ashore. It can also deliver messages to telex subscribers and to fax machines via a gateway into the telephone network.

Inmarsat-C is a 'store and forward' messaging system. You do not establish direct contact with the called party ashore and they cannot establish direct contact with you. You cannot therefore get into 'conversation' mode on your keyboard.

What happens is that the over-the-air system operates at the maximum 600 bps for your contact with the CES, via the satellite. Your message is stored in a 'buffer' at the shore end until the message is complete, after which the addressee is called automatically and the message passed.

One advantage of this system is that you do not have to wait for your addressee to be free before sending your message. Air time is therefore used to the best advantage, as users are not trying to establish the same call over and over again.

Another is the ability to send messages into a variety of different systems ashore — including telex, X400 messaging systems, computer terminals and fax machines.

In the to-ship direction, messages for you are also stored ashore until complete. The system will then call your vessel and once contact is established will pass the message on.

A 'mailbox' facility means that you do not have to leave your unit on all the time. You can conserve power by switching your Inmarsat-C kit off when you do not require it yourself, logging in occasionally to collect any messages from the mailbox. If you do not subscribe to a mailbox facility you need to keep your terminal logged in to the Ocean Region of your choice if you wish to receive messages from ashore.

If you communicate regularly with particular offices ashore, you need to ensure that they know which ocean region you will be monitoring, to ensure that your message is sent to an appropriate station for onward delivery.

SafetyNET

A big plus for Inmarsat-C is the SafetyNET facility. This is a one-way, shore-to-ship broadcast covering all sixteen Navareas for weather and navigational information(Fig 7.5).

A Inmarsat-C unit with a Global Positioning System (GPS) option included, can automatically select the broadcasts covering the Navarea you are operating in and reject all others. You will usually have a manual override which will allow you to select other, (eg adjacent areas) if you require.

The SafetyNET facility is part of the GMDSS infrastructure and covers the high seas in the same way that NAVTEX covers national waters.

Inmarsat-C Equipment

The above-decks antenna is the smallest Satcom antenna for any two-way system. Unlike the A/B/M antenna, the Inmarsat-C rig is omni-directional. There is no need for a gyro connection, nor any other satellite tracking facility. The antenna is very compact and can easily be sited on a sailboat mast, deck housing or antenna stub mast. Some units combine the GPS antenna with the Inmarsat-C antenna.

The below-decks equipment is about the same size as a VHF transceiver but, being a messaging system, it does need the addition of Data Terminal Equipment (DTE — a laptop computer is a common choice) and a printer.

Data Terminal Equipment and Printers

Equipment provided as part of a compulsory GMDSS fitting (eg, for passenger, trading and other commercial vessels) must have dedicated DTE and printing facilities.

If you are fitting Inmarsat-C on a private leisure craft, or other voluntary-fitted vessel, the DTE can be replaced by a laptop PC or other suitable device. Also, the laptop can be shared with your Marine SSB (for weatherfax, NAVTEX etc). It does not have to be a dedicated facility. Owners of voluntary-fitted vessels also have the choice of providing a printer, or not, as they wish.

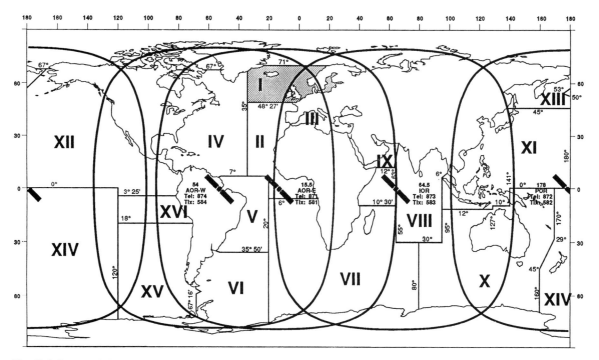

Fig 7.5 SafetyNET message to all ships in Navarea I (shown shaded). Standard-C terminals, EGC receivers and suitable A/B terminals can receive SafetyNET messages for their own (and adjacent) Navareas by programming their unit with the appropriate information. *(Courtesy of Inmarsat)*

Inmarsat-C with integral GPS or with a GPS interface, can be used in an emergency to transmit a distress message with your position accurately included should the need ever arise. If time permits, you can include any additional information which might help the authorities in the search and rescue operation.

FleetNET
The FleetNET facility allows owners of more than one vessel to broadcast 'fleet' messages to all, simultaneously. The messages are not recognised by other Inmarsat-C units monitoring the satellite at the same time.

The FleetNET facility could also be used by a group of owners with an interest in receiving similar information on a regular basis, from a single source ashore (eg, a news or financial information provider).

Prices for a Inmarsat-C fitting fall into the mid-price range of Marine SSB units - dearer than the smaller, more compact SSB fittings but cheaper than the commercial ship units often fitted on larger yachts.

Inmarsat-C and the Ocean Racer

The Inmarsat-C FleetNET facility was used during the 1993/94 Whitbread Round-the-World Yacht Race to exchange weather information and daily position reports with all participants.

Inmarsat-C was a compulsory communications fitting which, with a built-in Global Position System receiver, automatically updated race headquarters with the position of each boat. HQ would then broadcast the position of all yachts, to all listeners (bet that kept them on their toes!)

The weather information was a five-day forecast, broadcast to all participants in the normal 600bps format — which on-board computer terminals then decoded and reproduced as a weather map. The skippers were then able to choose a route to avoid the worst of the weather or catch the best winds.

EGC Receivers

In the same way that you can use a broadcast receiver in the MF/HF bands to receive weather and navigational information, you can use an 'Enhanced Group Call' receiver in the satellite bands.

EGC receivers will capture weather, navigation and other safety information broadcasts over the safety channel for the Navarea in which you are operating (or any other Navarea(s) you care to select). The EGC service is similar to the NAVTEX system, but covers the high seas navigation areas (Fig 7.5) rather than national waters.

L-Band EPIRB

The latest EPIRB system to come into operation is the L-Band service provided through Inmarsat.

This is seen as an alternative to the Cospas-Sarsat 406MHz EPIRB (see next Chapter) and has one distinct advantage in that it can provide an *instantaneous* alert, which *includes* your position (obtained from a GPS unit) directly to the rescue authorities ashore. There is no satellite nor shore processing of information required to work out your position.

EPIRB's are purely for distress alerting — they do not provide two-way communication, neither for commercial nor for emergency communication. Their prices are reasonably modest compared with most marine radio/Satcom kit and satellite EPIRB's do provide blue-water sailors with a way to notify the authorities ashore of any sudden calamity.

If the disaster which strikes your own craft leaves you in a life-raft or in the water in a remote location, with no other form of radio communication — a satellite EPIRB probably offers your best chance of rescue.

The L-Band EPIRB does not provide full global coverage (the polar regions are excluded) and any dependence on an external GPS may result in an out-of-date position being included in your distress call. Be sure to check the capabilities of any unit you are considering fitting.

Search and Rescue Transponders

L-Band EPIRB's may also be fitted with a Search and Rescue Transponder. The SART is a small transmitting device which operates on the 9GHz ship and aircraft radar band.

The SART will pick-up a pulse from a ship/aircraft radar, when in range, and will then transmit on that band. This pulse shows up on the rescue craft's radar as a straight line on the bearing of the SART.

SART's can also be fitted as a stand alone device, separately from the EPIRB.

Project 21

Project 21 is the Inmarsat organisation's thrust into future, personal, mobile telephone communications.

The idea of a single telephone number which can follow the businessman around his office and between offices has been around for some time. Project 21 is the mariner's answer to universal mobile, personal telephone communications, regardless of which vessel you are on or where that vessel may be.

As the title suggests, the realisation of this ideal is in the coming century, not too far off now, but neither does it provide us with today's answer. Something else to look forward to and to complicate today's decisions!

Installing and Commissioning Satcom Equipment

Inmarsat-A, B and M equipment requires a directional antenna system which should be installed and set up by a technician. Inmarsat-C and EGC receiver antennas are omni-directional, much smaller than their voice communication counterparts and are consequently easier to site and install. (Inmarsat produce 'Design and Installation Guide-lines' for each type of terminal.)

Any Satcom antenna should be situated clear of obstructions. Any superstructure (or person) which comes between the antenna and the target satellite can form a 'shadow' which can adversely affect the performance of the unit.

Each Satcom unit (A/B/M/C) comes with a commissioning application form which has to be completed by the installer and sent to your national 'Routing Organisation' — who will allocate an Inmarsat ID and arrange for commissioning tests to be carried out by your chosen CES. When the commissioning checks have been successfully completed the terminal is allowed into commercial service.

Maritime Mobile Equipment — Identity and Call Routing

Ship Earth Station Identification

Each Inmarsat SES is allocated a unique *Inmarsat ID*, so that calls from shore can be connected automatically to the correct terminal and mobile callers can be recognised and charged for calls made. The Inmarsat ID is derived from the Maritime Mobile Service Identity (MMSI).

The MMSI is the 'automatic connection' equivalent of the ships radio call-sign, used for terrestrial radio services (voice and telegraph). Your licensing authority will allocate an MMSI to your craft on request.

Armed with your MMSI and satellite equipment details, you apply to your national *Routing Organisation* (RO) for allocation of the Inmarsat ID and for commissioning of your new terminal. (Your mobile terminal supplier will do the installation and commissioning test, and will probably also complete the necessary paperwork on your behalf.)

The MMSI itself comprises nine digits — where the first three identify the country of registration, the next three digits identify the individual vessel (for that country), and the last three — 000 — signify that the vessel requires automatic connection into shore-based systems (DSC Alerting; Inmarsat telephone/telex networks etc). eg

    232 = UK register; 306 = USA register.

    xxx = individual UK/USA vessel.

    000 = 'access required to automatic systems'.

The RO will retain the first six MMSI digits as part of the new Inmarsat ID, but will adopt it as follows.

A new additional first digit is added to identify the terminal type as M, C, B or A — where

    6 = an M-terminal

    4 = a C-terminal

    3 = a B-terminal

    1 = an A-terminal

The final zero's are replaced by two digits, to identify an individual terminal or service port on the same terminal onboard your vessel. The ID therefore consists of 9 digits —

eg.(1) 6 = M-Terminal

    232 = UK registered vessel

    (xxx) = (individual UK registered craft)

    10 = Telephone port on the M-terminal.

eg.(2) 6 = M-terminal

    306 = USA registered vessel

    (xxx) = (individual USA registered craft)

    11 = Fax port on the M-terminal.

(NOTE: The common numbering system based on the MMSI was not used for the original Inmarsat-A service. Not all A-terminal IDs will follow the above pattern.)

Calls from Shore-to-Ship

When the shore caller wants to speak to a mobile (or wants to send a telex to a mobile), they need to know which Ocean Region (which of the four satellites) on which the vessel is keeping watch and the individual Inmarsat ID. It is then just a case of making an international telephone/telex call — eg.

> A call from a telephone in the UK to a vessel fitted with M-Sat, cruising off the east coast of Australia:
> 010   (International call prefix from the UK)
> 872   (Pacific Ocean Region — telephone prefix)
> 6232xxx10   The Inmarsat ID for an 'M' terminal of the vessel being called.

International call prefixes differ from country to country and will be listed in front of your telephone/telex directory. The Ocean Region codes remain the same, regardless of the country from which the call is being made.

Calls from Ship

Making a call from ship requires the user to select a particular satellite *and* an individual Coast Earth Station. You need to consider the following points when choosing either.

Are you in 'sight' of more than one bird? For example — the North Sea between the UK and mainland Europe is covered by three of the four Inmarsat satellites, AOR-W, AOR-E and IOR. You can therefore choose to communicate through any of the three.

But not all CESs in a particular Ocean Region support all services (M, C, B, A) or provide the same range of facilities within the service type supported.

Some terminals come with a 'default' CES, for each Ocean Region, pre-programmed into the terminal software. As a buyer, you need to give some though as to which default CES you would prefer to have programmed in, especially for a unit which is not easily re-programmed by the user. The default CES *can* be over-ridden by the user, but the wrong choice at the outset could prove to be a nuisance.

An automatic telephone call from a vessel in the Caribbean Sea, through Goonhilly CES, to an office in Singapore, would follow this typical process:

(Assuming an M-terminal, with Southbury CES programmed as default)

Key	Action
	Lift the handset and hear dial tone.
	(Your SES should display 'Ready'.)
(See Note 1)	Select AOR-W. (Your terminal will now communicate through the default CES for AOR-W, unless you ask for another.)
*002	Selects Goonhilly CES (overriding default). (See Note 2)
00	Select automatic connection.
56	Country code for Singapore.
4609200	The telephone number of the called party.
#	Hash sign is the 'send' button — your call details will now be sent via the satellite to Goonhilly for connection.

Note 1
Follow manufacturers instructions for 'how to select an Ocean Region'.

Note 2
If you want to communicate through the default CES, you should omit this step.
The * before the number is needed to tell your terminal that a CES ID follows.
(This is the correct procedure for a Scientific Atlanta 'Maristar-M' terminal — other manufacturers may adopt a different approach.)

Antenna Alignment

Most Satcom antennas (except Inmarsat-C units) have to be aligned with a satellite upon installation. The antenna may also have to be re-aligned when the unit has been switched off for a time — especially if you have been swinging around a buoy.

The alignment is a combination of azimuth (bearing) and elevation (from the horizontal). Antenna alignment details are included as Appendix P, to help you to re-align your antenna whenever necessary.

Summary

There are three types of Satcom unit providing voice telephone facilities — Inmarsat-A, B and M. Inmarsat-B and M are the most recent and they offer the most economical call charges. They are the ones to choose from if you want Satcom telephone communication.

Inmarsat-B is for larger vessels and Inmarsat-M — depending on which manufacturers unit you choose — can be fitted on all but the smallest of craft.

Both B and M provide fax connections in addition to voice, and either can provide data communications, although at different speeds. Priority communications in the event of an emergency are also available on all Inmarsat-B and Inmarsat-M units. Inmarsat-B can send and receive telex messages, Inmarsat-M will not.

Inmarsat-C is a two-way, store and forward messaging system. Compact (both below and above-decks), it also provides a distress alerting and distress messaging facility. Inmarsat-C units can receive all SafetyNET and FleetNET broadcasts.

EGC receivers are one-way, shore-to-ship units which will provide you with weather and safety information (SafetyNET) and will also alert you to distress operations in your area.

The L-Band EPIRB offers an alternative to the 406MHz Cospas-Sarsat EPIRB, which is described in the next Chapter. L-Band EPIRB's, like the other Inmarsat services, are not effective in the polar regions (should you be considering going there!), and some units may not include a fully up-to-the-minute position in a distress situation.

Satcom equipment is generally more expensive than terrestrial radio counterparts but does, for the most part, provide easier and more timely connections over high quality links and with the minimum of training.

Chapter 8
Other Satellite Systems

Introduction

In addition to the Inmarsat system, which offers a two-way communication service for commercial messages and distress and safety messages, there are four other communication satellite systems of interest to mariners. Two are already in operation; the other two will offer alternative commercial services to those provided by Inmarsat.

The first (and most important to the mariner) is the Cospas-Sarsat system, which will process a distress alert from a compatible (ie 406MHz) EPIRB. The second forms the Amateur Satellite (AMSAT) system, and is for radio experimental purposes only. The two proposed commercial services are known respectively as Iridium and Odyssey, and are planned to come into operation towards the end of the 20th Century.

The Cospas-Sarsat System

If the near 70-nation Inmarsat co-operative is unique, then the Cospas-Sarsat partnership would have raised a few eyebrows had it received the level of publicity given to other 'East/West' engagements of the 1980's. Whilst the United States of America and the former Soviet Union were arguing for and against a 'star-wars' military satellite system, they got together with a handful of other countries to provide a satellite constellation which now forms the Cospas-Sarsat system — one which can pin-point a casualty to within a few miles, even in the remotest of areas, and which now forms an integral part of our Global Maritime Distress and Safety System.

System Origins

The Cospas-Sarsat system was instigated by Canada, who sought a way of locating downed aircraft in the vast areas of mountain and forest which cover so much of their country. The Canadians were able to demonstrate the feasibility of satellite detection of 'Emergency Locator Transmitters' (ELTs, as the aircraft/land units are called) and were joined by France and the USA to start the development of the system.

The USA's National Oceanic and Atmospheric Administration (NOAA), NASA, and the Centre National d'Etudes Spatiales (CNES) of France were already using satellites to collect data from space, including weather information, and they came into partnership with Canada to test a system for tracking and locating ELT's. By the end of 1979 the Soviets, who were developing their own 'COSPAS' system (meaning 'Space System to Search for Distressed Vessels') joined the partnership, which has resulted in the shared satellite constellation and ground infrastructure which now covers all of the earth's surface, both land and sea.

Present System

The present system consists of a payload on each of four satellites at any one time, two COSPAS birds and two 'Search And Rescue Satellite Aiding Tracking', or SARSAT transponders, the latter being carried as part-payload on NOAA satellites of the USA.

The Cospas-Sarsat satellites are polar-orbiting birds, not the geostationary type used by Inmarsat. Each Cospas-Sarsat satellite orbits the earth on a North/South, South/North path, passing over both poles (Fig 8.1).

The satellites orbit at a much lower level (about 530 miles for the SARSAT and 620 miles for the COSPAS birds) than their Inmarsat counterpart and between them cover the entire surface of the earth every few hours, monitoring the 121.5MHz and 406MHz EPIRB/ELT frequencies. (The NOAA units also monitor the military aircraft frequency of 243.0MHz).

Early tests had indicated that signals captured on 121.5MHz could produce a position accuracy to around 12 miles/20km which, at first glance, looks quite good. But that level of accuracy would require a visual search pattern of many hours, even in the open sea. To help find a downed aircraft in the wooded, mountainous Canadian wilderness required something much more accurate. Additionally, with over 350,000 unregistered beacons in service, the potential cost in abortive SAR missions due to false alarms was unacceptably high.

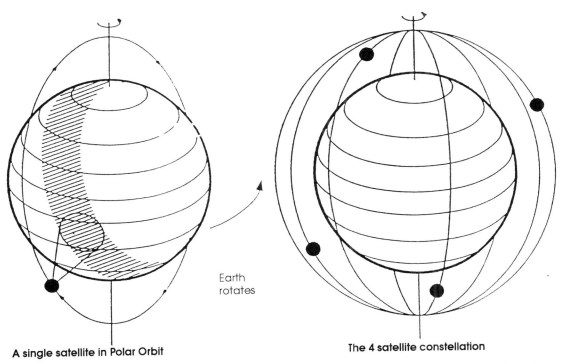

Earth rotates

A single satellite in Polar Orbit

The 4 satellite constellation

Fig 8.1 Cospas-Sarsat Polar Orbiting Satellites. There are normally four operational satellites deployed at any one time — two Cospas and two NOAA/Sarsat birds.

Personal Locator Beacons

Small EPIRB's operating on 121.5MHz are being marketed as Personal Locator Beacons. These are ideal as additional beacons in the event of someone falling overboard.

The PLB can be automatically or manually activated and will activate an alarm device, if fitted, onboard your craft. You can also fit a 121.5MHz 'Homing Receiver' onboard. (If you ever find yourself in the sea with one of these beacons strapped to your upper arm, then please remember two things: (i) You are now at sea level. The visual horizon is very limited. (ii) If your beacon antenna is just below, rather than above the sea surface, your signal will be lost).

As long as you know that someone has gone overboard, you can alert the rescue authorities and provide them with the approximate position. The PLB will then allow a rescue vessel/aircraft to home in on the individual.

Due to the very large number of 121.5MHz beacons already issued, rescue authorities are unlikely to instigate a SAR operation based solely on the receipt of a 121.5MHz signal. They will normally want confirmation that an aircraft, ship or person is missing before instigating an expensive operation.

406MHz EPIRB

The modern 406MHz (actually, 406.025MHz) EPIRB/ELT, combined with the Cospas-Sarsat tracking system, will plot a position to within 1—3 miles (2—5km) resulting in a visual search area which a SAR aircraft could cover in less than one hour, once on-scene. Considerably more promising on land and, for seafarers, possibly the difference between rescue alive and (reasonably) well, or death by hypothermia.

Satellite Tracking System — 406MHz

The satellites are constantly monitoring 406.025MHz as they orbit the Earth. They also monitor 121.5MHz.

Only the 406MHz signal will allow the satellite to plot the position to the desired degree of accuracy on a single pass and this is done by taking advantage of phenomenon known as the 'Doppler effect'. Doppler shift is an apparent change in the frequency of a signal, or the pitch of a sound, when heard at a particular location, where the transmitter and/or receivers are on the move. The 'shift' is demonstrated on Fig 8.2

If a constant sound (or a radio transmitter at a constant frequency) is travelling rapidly towards the 'receiver' (eg. your ear or a radio receiver), then the pitch/frequency will appear to be higher than the actual source frequency. As the source passes close by the receiver there is a sudden drop in pitch/frequency, followed by a steady signal at a frequency *below* that of the source.

One *sound* example is that of a person standing at the edge of a motor-race track, as we have all experienced in the movies, if not in real life! As the car approaches we hear a steady engine noise (getting louder but at a steady pitch). When the car passes our position there is a sudden drop in pitch, followed by a steady tone at a noticeably lower level. The *volume* decreases as the car moves away but the *pitch*, again, sounds steady.

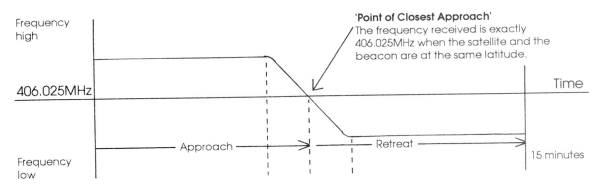

8.2a Frequency shift is at its greatest, resulting in a sharp drop, where the longitude of the satellite ground path is close to the longitude of the EPIRB.

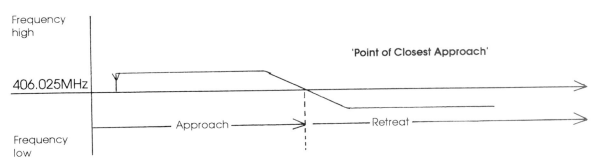

8.2b Frequency shift is at its narrowest where the EPIRB is on the horizon of the satellite's vision. The signal is also heard for a lesser period.

The 406MHz beacon transmits a 5 Watt radio frequency signal burst of approximately 0.5 seconds duration, every 50 seconds. The satellite needs to capture at least four of these bursts, which are superimposed upon each other to obtain a usable signal. Where the beacon is in sight for 10 minutes the bird 'sees' $\frac{10 \times 60}{50} = 12$ bursts.

For a sighting of 15 minutes the satellite will see $\frac{15 \times 60}{50} = 18$ bursts.

Fig 8.2 Doppler Shift

The same thing happens with a radio frequency signal, and it does not matter whether it is the transmitter which is moving or the receiver. The radio frequency will be high and will get stronger on the approach and will drop through, and then below the true frequency, getting weaker (but at a reasonably steady frequency) as it moves away (retreats).

As our Cospas-Sarsat satellite orbits the Earth, it is on the look-out for a stationary transmitter (our EPIRB). When the bird picks up a signal it measures the received frequency (which will be 406.025MHz *plus* an amount caused by the Doppler effect). This is shown on Fig 8.2 'approach'.

As the bird gets close to the latitude of the EPIRB there is a marked drop in frequency and the drop passes through the true 406.025MHz mark as the satellite passes over the same latitude as the EPIRB. This is the 'Point of Closest Approach'.

The drop continues for a short time and like our passing racing car, settles down to a steady frequency, somewhere below 406.025MHz, until the satellite passes out of view of the EPIRB.

Depending on the relative longitude of the satellite and the EPIRB, the signal will be heard for about 10—15 minutes. Should a satellite pass directly over an EPIRB, the beacon will remain in sight for the longest time and the high/drop/low pattern will be very pronounced (Fig 8.2(a)). An EPIRB on the horizon will remain in sight for a shorter period, and will produce a less pronounced shift (Fig 8.2(b)). The satellite is able to compute the *distance* of the EPIRB from the satellite's own ground track, according to the amount of shift.

On its own, that calculation would provide two possible positions, one to the East and one to the West of the satellite ground track, but on the same latitude.

There is a second Doppler shift involved, that caused by the rotation of the Earth. The Cospas-Sarsat birds are able to use this additional shift to determine whether the EPIRB is East of the satellite ground track, or West (Fig 8.3).

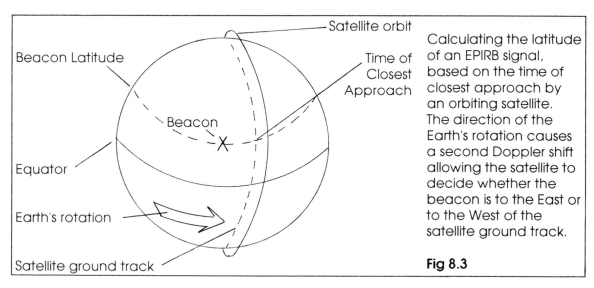

Calculating the latitude of an EPIRB signal, based on the time of closest approach by an orbiting satellite. The direction of the Earth's rotation causes a second Doppler shift allowing the satellite to decide whether the beacon is to the East or to the West of the satellite ground track.

Fig 8.3

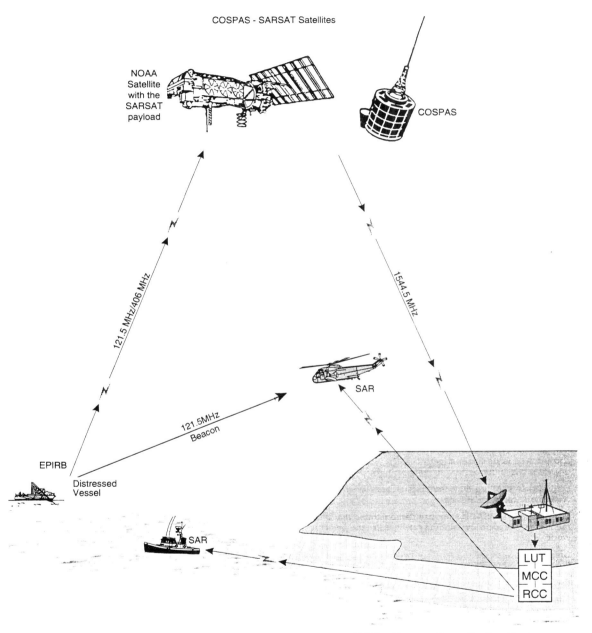

COSPAS - SARSAT Satellites

NOAA Satellite with the SARSAT payload

COSPAS

121.5 MHz/406 MHz

1544.5 MHz

SAR

121.5MHz Beacon

EPIRB

Distressed Vessel

SAR

LUT
MCC
RCC

LUT = Local User Terminal   MCC = Mission Control Centre   RCC = Rescue Co-ordination Centre

Fig 8.4 The COSPAS-SARSAT System for Mariners *(Courtesy of COSPAS-SARSAT)*

Once the satellite has received and processed the EPIRB signal, it needs to pass it on to the authorities ashore. The system for achieving this is demonstrated in Fig 8.4.

A satellite picks up an alert and carries out the signal processing, extracting identification, recording the time that the signal was exactly 406.025MHz, and recording the effects of the two Doppler shifts. This information is then broadcast on a downlink frequency (in the 1.5GHz band) to any LUT in sight, in 'real-time'. The information is also stored onboard until the satellite can pass it directly to the next LUT which it passes over, if not in real-time contact. (LUT area coverage shown in Fig 8.5).

The LUT carries-out the final processing of the information received, to determine the location of the EPIRB and passes information to the Mission Control Centre.

The MCC, using the position of the casualty, decides which Rescue Co-ordination Centre is best placed to handle the Search and Rescue (SAR) operation. This will depend on where the casualty is located and which country has SAR jurisdiction for the waters concerned. The RCC will co-ordinate SAR forces who, hopefully, will recover the distressed persons and the EPIRB.

Homing-in on the Casualty

Most 406MHz EPIRBs also broadcast on 121.5MHz, the civil aircraft distress frequency. This is to allow SAR aircraft to home-in on the exact position of the EPIRB (SAR aircraft carry a homing device which works on 121.5MHz).

It is a good idea, if you have to abandon ship, to secure the EPIRB to your survival craft with a lanyard. This will ensure the quickest possible detection and the recovery of people and of EPIRB. When rescue is at hand the EPIRB can be switched off, preventing repeated and unnecessary activation of satellite systems.

EPIRB's should not be kept inside a covered liferaft when in use, as liferaft covers can considerably attenuate your transmission, reducing the likelihood of detection.

Global Coverage

The Cospas-Sarsat system provides total global coverage with its polar orbiting satellites. If a LUT is in sight at the same time of the beacon signal being received, it will take around half-an-hour for message processing and alerting of authorities.

Not all sea-areas are in sight at any one time however, and signals picked up when the satellite is outside the coverage area of a LUT must wait to be passed ashore. If you become a casualty in an area which is far from an LUT station, it can take up to three hours before the LUT receives your data and passes it on to the MCC/RCC. The SAR operation will then move into action.

EPIRB Registration

One of the benefits of the 406MHz EPIRB against earlier types is that a system of registration has been put in place. This allows the RCC to check possible false alarms with the registered owner, before mounting a costly SAR operation. It also logs the type and description of the craft, so SAR units will know what to look for (much time can be wasted trying to attract the attention of vessels in the general area of a casualty, only to find that

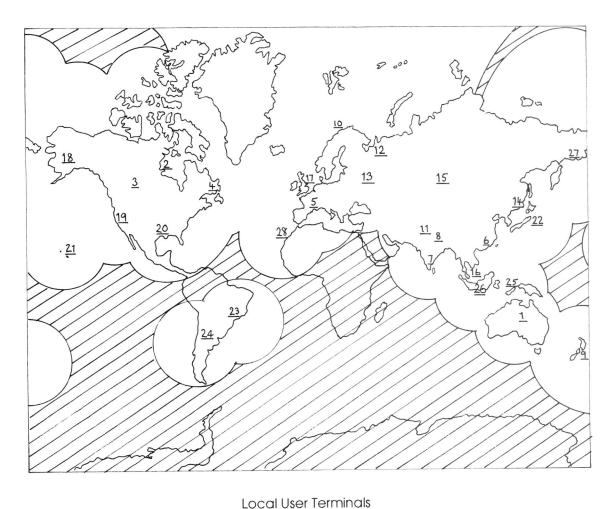

Local User Terminals

No	LUT				
1	Alice Springs, Australia	10	Tromsoe, Norway	20	Texas, USA
2	Churchill, Canada	11	Lahore, Pakistan	21	Hawaii, USA
3	Edmonton, Canada	12	Arkhangelsk, Russia'	22	Yokohama, Japan
4	Goose Bay, Canada	13	Moscow, Russia	23	Brasilia, Brazil
5	Toulouse, France	14	Nakhodka, Russia	24	Santiago, Chile
6	Hong Kong	15	Novosibirsk, Russia	25	Ambon, Indonesia
7	Bangalore, India	16	Singapore	26	Jakarta, Indonesia
8	Lucknow, India	17	Lasham, UK	27	Tilichiki, Russia
9	Wellington, NZ	18	Alaska, USA	28	Maspalomas, Spain
		19	California, USA		

The map represents real-time coverage of the COSPAS-SARSAT constellation. A 121.5MHz EPIRB alert will only be relayed if both EPIRB and LUT are in sight of the satellite at the same time. 406MHz alerts will be stored aboard the bird until a LUT comes into view.

Fig 8.5 Real-time Coverage of COSPAS-SARSAT Constellation.

they do not know what is going on, but that they themselves are not in distress, 'thank you very much'). It is much better from the outset if SAR units know they are searching for a 30ft yacht, a 60ft fishing vessel or a container ship.

Automatic/Manual Activation

EPIRB's have been designed for automatic and/or manual operation.

Automatic activation of maritime EPIRB's is normally achieved by a 'hydrostatic' process. The unit is mounted on a special bracket and in a clear position above-decks (eg, on the mast, guard-rail or cabin top).

The hydrostatic system reacts to water pressure and will automatically release the EPIRB when it reaches a pre-determined depth. This might not be much help if you capsize but stay afloat, or if your wooden mast comes adrift and floats away with the EPIRB attached. A dual-action unit is therefore preferable.

Unfortunately, there is no single answer to suit all circumstances — you just need to be sure to fit an automatic device where it is not likely to be fouled — but preferably where you can get hold of it yourself if you have to abandon (eg, in the event of a fire onboard?)

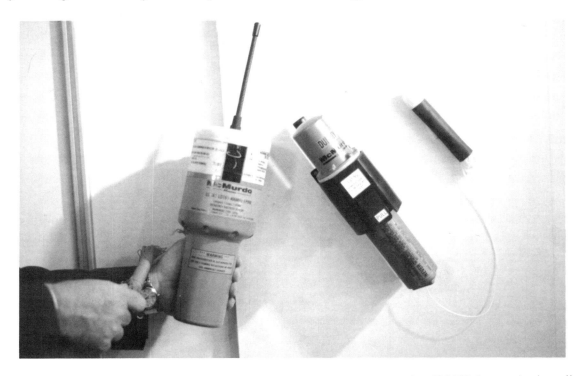

The EPIRB (being held) and the Search and Rescue Transponder (SART) (mounted on the wall) both have lanyards attached — the lanyard should be secured to the life-raft or otherwise kept in the vicinity of the casualty.

Testing Your EPIRB

EPIRB's (the modern 406MHz type) will normally have a battery life of 4—6 years. Despite this long battery life, you should carry out the periodic 'self-test', according to the manufacturers instructions. This will test the internal circuitry of the EPIRB without activating the device. The self-test will indicate whether the unit requires servicing or battery replacement.

Should you purchase a vessel with a 406MHz EPIRB already fitted, you should carry out the self-test procedure and inform the relevant authority of the change of name and contact number for the new registered owner.

EPIRB's, like other marine electronics, need to be secure from theft. Some units have a lockable enclosure for this purpose. If you fit a locking unit to your own boat, make sure the key is attached to the same ring as your ignition key so that it is never left at home — and never leave harbour without first unlocking your EPIRB case.

An alternative is to take the EPIRB home with you on each occasion you leave the boat, but again — make sure you take it with you on your next trip!

Type of Distress?

Finally, some units have a facility for including a distress 'type' in the transmitter signal. This is achieved by manually selecting a number on the unit, to correspond with the trouble in which you find yourself.

The standard types of distress which can be indicated by such units are:

1	Fire or Explosion
2	Flooding
3	Collision
4	Grounding
5	Listing or Capsizing
6	Sinking
7	Disabled and Adrift
8	Abandoning

As you can see, there is a variety of circumstances which would not result in automatic release of the EPIRB, and some which could cause your unit to be destroyed if you do not keep hold of it.

Amateur Satellite Systems

The other existing satellite system which will be of interest to *some* mariners is the Amateur Satellite System.

Amateur radio facilities and their use vary from country to country. In the USA the very practical 'phone-patch' offered on the HF bands (Chapter 5) is at one end of the spectrum, whilst the experimental and educational use of the amateur satellites is at the other.

Radio amateurs are, by international definition, 'experimenters'. For those who spend time at sea therefore, an opportunity to experiment in a different environment and from various locations presents itself. Amateur satellites can add to the interest.

A number of different satellites and satellite systems operate on amateur frequencies and, like the Cospas-Sarsat initiative, they bridge international and cultural gaps by encouraging multinational usage and inter-nation working.

Current amateur satellites (mostly called 'OSCAR' followed by a number — eg Oscar 11, Oscar 12) have been paid for by amateur radio enthusiasts in a number of countries and can only be accessed by licensed radio amateurs.

Only the most enthusiastic maritime mobile radio ham is likely to try to track an amateur satellite whilst at sea, with the need for an antenna which is steerable in both elevation and azimuth — particularly as the OSCARS operate in different frequency bands that require a separate antenna for each.

If you are interested in pursuing this aspect of amateur radio as a hobby (it should never be seen as an *alternative* to a maritime safety system), you can obtain details of your own national amateur satellite organisation (AMSAT-UK, AMSAT-USA etc) through your national Amateur Radio Society.

Iridium™ Satellite Phones

Mobile Satellite Service Frequency Allocations
A 'World Administrative Radio Conference' (WARC) in 1992 made significant changes to the allocation of radio frequencies for 'Mobile Satellite Services'.

On the one hand, frequencies which were previously allocated for maritime use, *exclusively*, were opened-up to 'generic' use in ITU Regions 2 and 3. On the other hand, a range of additional frequencies was provided for generic use. As 'generic' includes mariners, we can now have access to more frequencies than before.

The change is a double-edged sword. 'Generic' use means that land-mobile users (who far outnumber maritime users) will have equal access to the frequencies. But the resulting increase in the size of the 'market' will result in lower equipment costs and lower call charges — something mariners have desired for many years.

Another effect is that new service providers are positioning themselves to take a share of the much larger market. Two proposed new systems, which will compete with a MSS offering from Inmarsat, are described below.

The Iridium Communication System
The Iridium system infrastructure, like that of the present Inmarsat system, will be based on a satellite constellation which is operated and maintained by one organisation ('SATCOM' — Motorola's Satellite Communication Division); and land-based gateways into national and international telecommunications networks — those Gateways being operated by telecomms service providers in the Region(s) concerned.

In addition to the satellite constellation and the Gateways there will be a System Control facility, and a variety of mobile terminals (hand-helds; vehicle/boat telephone units; and pagers are envisaged).

Satellite Constellation

The Satellite Constellation will consist of sixty-six satellites in six planes (ie, eleven birds in each plane) — in a polar 'Low Earth Orbit' (LEO) — See Fig 8.6. The orbital planes will be spaced at 31.6° at the equator — except for the unavoidable 'counter-rotational seam' where spacing is reduced to 22° to aid inter-satellite communication. Between them, the satellites will provide total global coverage.

The LEO height for Iridium is to be 420Nm — a bit lower than that of the Cospas-Sarsat birds. Like Cospas-Sarsat, however, the speed of the satellites over the Earth's surface will cause the communication frequencies to alter (Doppler-shift) — and the satellite system will have to be sophisticated enough to cater for this change *without* distorting the message. (One system's advantage is another's burden!)

An advantage of LEO is the minimal delay between one party stopping talking and the other hearing what was said, something which new users of Geo-stationary systems tend to find disconcerting. Removing the perceptible delay will make conversation more natural.

Each satellite will carry three antennas for communicating with mobile terminals. Each antenna will use sixteen spot-beams, making a cellular type honeycomb of forty-eight cells from each satellite. This allows considerable re-use of the same frequencies within different (non-adjacent) beams, something which is essential if frequency and system congestion is to be avoided. (The outer beams will be turned-off as the birds approach the poles, thus avoiding overlapping coverage, and conserving power).

Like cellular systems, users will be 'handed-off' between beams in the same satellite and, when required, from one satellite to the next. (As each satellite will orbit the Earth in around 100 minutes, hand-off will be commonplace).

Each Iridium satellite will have communication links to the bird immediately ahead and immediately behind on the same plane and up to four links with satellites on adjacent planes for inter-satellite hand-off.

System Control

Satellites are complicated beasts and they have to be kept on track, at the correct height and in the correct attitude (ie — with their antennas facing the area they have to communicate with). The Iridium system will use two geographically separated system control facilities to check and maintain the configuration of the sixty-six birds.

Telephone Network Gateways

Gateways will provide the link between the satellites and the Public Switched Telephone Networks of the host country — for from-mobile and to-mobile calls. Each Gateway will have three dish antennas, (each about ten feet in diameter) which will be steered to track the satellites. Antennas will be about twenty miles apart to reduce any likelihood of loss of communication-path during thunderstorms (more prevalent in some areas than others).

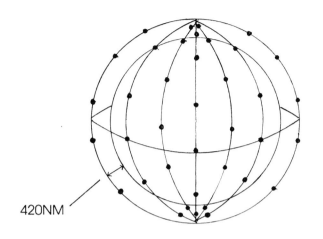

— Eleven satellites in each of six planes.
— Alternate planes have satellites at corresponding latitudes.
— Planes are spaced at 31.6° except at the unavoidable 'counter-rotational seam', where separation is reduced to 22°.

420NM

Fig 8.6 Planned Iridium Constellation

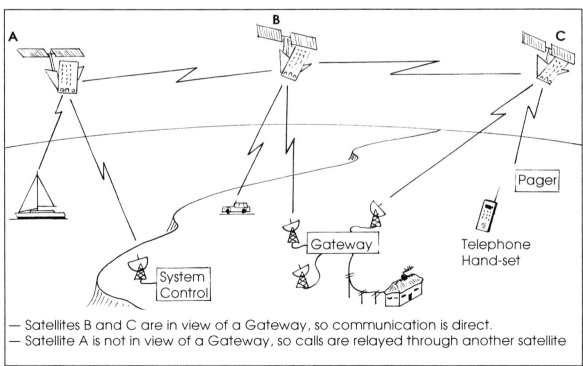

— Satellites B and C are in view of a Gateway, so communication is direct.
— Satellite A is not in view of a Gateway, so calls are relayed through another satellite

Fig 8.7 Iridium System Overview

Two of the dishes will be needed to track two satellites simultaneously — as one bird drops towards the horizon, another is coming into view. The third dish will act as a standby — eg, for maintenance periods.

Mobile Terminals — Cellular Compatibility

Telephone units for hand-held and vehicle/boat fitting will offer two methods of accessing the PSTN. The first (and automatically preferred method) will be through your 'home' cellular radio network; the fall-back (where your home cellular network is not in range) will be the Iridium system. This means that you will invariably be connected to the (less expensive) cellular network when in range, and that the Iridium channels will not be saturated by users who had such an alternative.

For mariners, the Iridium system will provide all telephone communication when out of home waters.

Odyssey™ Communications System

The final contender for interest in terms of Mobile Satellite Systems is the 'Odyssey Satellite-Based Communications System' from TRW Inc of California.

Like Inmarsat's proposed 'Global Personal Communication System' and Iridium — Odyssey is likely to come into service around 1998 and is expected to use combined Cellular/Satellite mobile units. At present TRW are talking only in terms of a 'hand-set' — not about mobile units specifically for road vehicles or seagoing craft.

The Odyssey satellite design will be based on an improved version of the US Navy's Fleet Satcom spacecraft, which TRW upgraded in the early 1990's.

Ironically, the new system will not provide 100% coverage of sea areas world-wide, as it is designed to provide service to the major land-masses. The coverage provided however, will be adequate for the vast majority of voluntary-fitted craft who remain in home waters.

TRW expects Odyssey to be the least expensive of the proposed new Mobile Satellite Services, as regards cost of *provision* of the service and, for that reason, expects to be in a position to offer the lowest call charges of the competing Mobile Satellite Services.

Satellite Constellation

The Odyssey satellite constellation (See Fig 8.8) is, like Iridium and unlike the present Inmarsat system, an orbital constellation. TRW have chosen a 'Medium Earth Orbit' (MEO) at 5,591 Nautical Miles high, rather than the LEO of Iridium.

The greater height of the Odyssey orbit allows global coverage with fewer satellites in fewer planes. Odyssey is currently planned to have twelve birds only — four in each of three orbital planes.

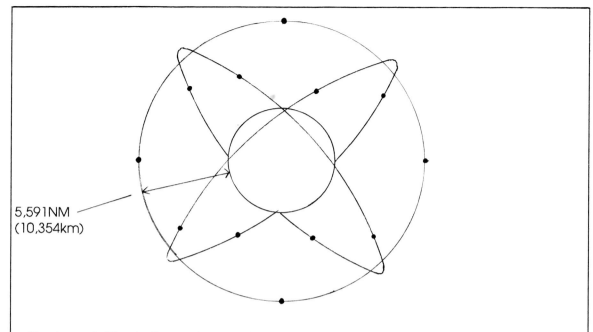

5,591NM
(10,354km)

— Twelve satellites in three planes. (Four birds in each plane.)
— Satellites placed in 'Medium Earth Orbit'. (Much higher then LEO but considerably less than GEO.)
— Satellite antennas pointed at main land-masses to cover maximum number of users with the optimum number of satellites.

Fig 8.8 Odyssey Satellite Constellation

Satellite Orbits for Mobile Satellite Services

The existing Inmarsat system (Inmarsat-2 satellites) comprises four geostationary satellites, orbiting above the equator at the same speed as the Earth's rotation, at a height of 22,200 miles (36,000km). Each satellite has one large 'beam', or footprint, covering its entire service area.

Inmarsat is due to launch its third generation of satellites during the second half of the 90's decade and this generation will have 'spot beam' capability.

Inmarsat will also be adding to their four geostationary satellites — to form an 'enhanced Geostationary' constellation — and the additional birds will probably be in LEO; or in MEO; or in 'Intermediary Circulatory Orbit' (Well — it would be boring if they all did the same thing, wouldn't it?). Inmarsat will then be in the position to provide their own 'Global Personal Communication System' with which Iridium and Odyssey will compete.

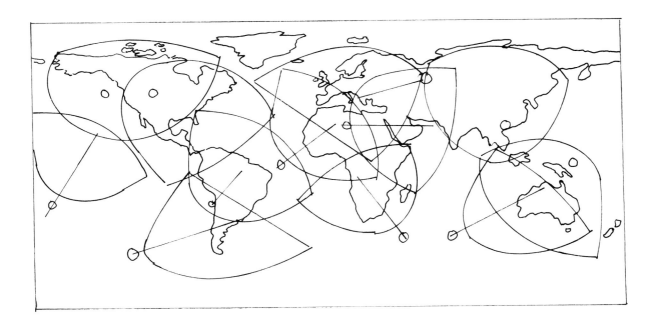

Fig 8.9 Odyssey World Coverage

Coverage

Although Odyssey will offer coverage all around the world, the system is not designed to provide *complete* coverage of all areas. In particular, large areas of the South Pacific and Indian Oceans, as well as Greenland and other polar regions, will not receive service unless present plans are amended. (Fig 8.9)

The reason for this is TRW's desire to provide simultaneous dual-satellite coverage for all the mainland areas, where the land mobile market will be concentrated, with the minimum number of satellites. They intend to achieve this using satellites which can have their antennas directed at a particular populated region when that region comes into view.

Each satellite will have its attitude altered to keep the antennas directed towards the service area as it passes across a region and moves away towards the pole or towards another region.

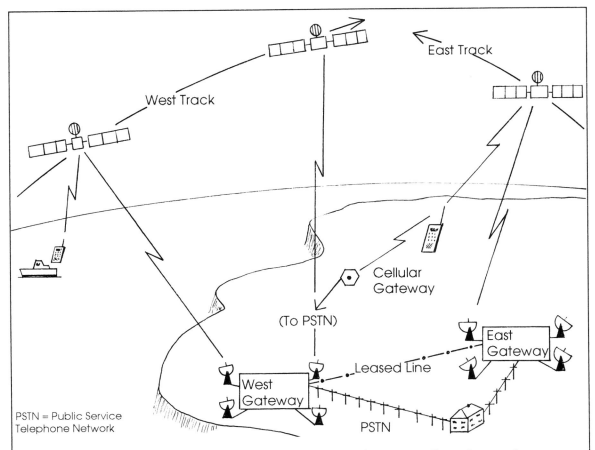

— With dual satellite coverage in each service area at any one time, two gateways are required.
— Each Gateway requires enough antennas to handle two satellites on the same plane simultaneously, sometimes a third, and also to provide diversity reception to avoid service disruption by bad weather.
— Hand-sets will access an appropriate cellular radio system if in range and an Odyssey satellite if not.

Fig 8.10 Odyssey System Overview

System Overview

With dual-satellite coverage planned for each of nine regions or 'service areas', the system is envisaged to have two 'Gateways' in each region (Fig 8.10).

Each Gateway will have four tracking antennas — each of ten feet diameter — placed about 19 miles/30KM apart. Two or three dishes may be tracking satellites at any one time, with the fourth being available to ensure seamless handover from one satellite to another and for diversity reception.

Diversity Reception

Diversity reception is used with some radio communication systems to help ensure continuous service. By spacing antennas geographically apart, the different antennas can receive the same signal — but one signal may be stronger than the other. Control equipment recognises the stronger signal and connects that antenna to the Gateway equipment.

Mobile Satellite Systems will be operating in frequency bands which are prone to disruption by thunderstorms. By separating the antennas for a single Gateway by around 20 miles/30km, it is expected that there will always be a path from satellite to Gateway around any thunderstorm.

As a satellite comes into view of a service region it will start to pick-up any new calls from that region. The preceding satellite will take on no new calls as the traffic on the new satellite builds up.

There will be a period of around ten minutes when a service area is in view of two satellites on the same plane, allowing most calls on the first satellite to be completed without handover.

Each satellite will remain in view of a service area for about two hours. The intention is to use each satellite for call handling when its angle of view is reasonably high (upwards of thirty degrees), so reducing the likelihood of shadowing by buildings or rough terrain.

There will be nineteen spot beams, or 'cells', within the main-beam coverage of an Odyssey satellite antenna. As each cell is about 500 miles/800km in diameter, and as the satellite attitude is adjusted as it passes over a service area (to keep the antenna orientated towards that area), the need for call handover between the cells within any one beam will also be minimal.

On the few occasions where a (long) call has to be passed from one satellite to another, the handover will be carried out at the Gateway without any action by the caller.

The two Gateways in any one region will be linked together by leased-line for control purposes.

Mobile Terminals (Hand-sets)

Odyssey hand-sets will be made by adding the Odyssey communication 'chipset' to existing cellular hand-sets. In particular, the system will be able to work with key cellular systems around the world. This is intended to include the Pan-European digital cellular system, which was opened for commercial use in 1993; and the Advanced Mobile Phone Service (AMPS) and networks using the American Digital Standard (ADS) in the USA. A user will then choose the cellular system required and the hand-set will cover that system, along with Odyssey.

As there will be cellular terminals on the market other than hand-sets, terminals more suited to road vehicle/boat fitting, the mariner will no doubt be offered a more appropriate installation.

TRW consider that their Odyssey proposal will provide the best overall solution for a

Mobile Satellite Service. The MEO is low enough to keep propagation delay to a level which allows 'natural' conversation but high enough to remove any need to compensate for Doppler shift.

By keeping the satellites as simple radio relay devices, they can handle different types of modulation if required. And each spacecraft can be reconfigured to increase the number of spot beams, increasing the number of communication channels available.

What's in a Name (1)?
The name Iridium was chosen because 'Iridium' is the name of the element with an atomic number of 77 — the number of birds in the original plan. When the proposal was changed to 66 satellites — 66 being the atomic number of the element Dysprosium — prudence suggested the retention of the original name. After all, they want to sell a personal communication service, not a laxative.

If Motorola and friends ever considered retaining the seven-plane constellation and dropping one bird in each plane — for a total of seventy satellites — a glance at the Elemental Atomic Number table will reveal why that particular proposal would also have retained the name Iridium!

What's in a Name (2)?
As Motorola and others are considering moving into the maritime communications market, so Inmarsat has spread its wings to cover aeronautical and land-mobile communications. The organisational name of 'International Maritime Satellite Organisation' is also under review, in an attempt to come up with a name more suited to the wider role which Inmarsat has chosen to take on.

However, the Inmarsat commitment to the maritime market — and particularly to global distress and safety communications — is expected to remain strong.

The overall effect of the changes generated by newcomers to the maritime market and of Inmarsat (or In-Mob-Sat, if that becomes their choice) widening its own customer base to serve new markets, will hopefully further increase the choice for mariners, with reductions in equipment costs and call charges, without prejudicing safety services.

'Watch', as they say, 'this space'.

Summary

The amateur satellite system, based mainly on OSCAR satellites, is for experimenting and education purposes. Whilst the system will test the ingenuity of maritime mobile amateur radio enthusiasts, it offers neither commercial communications nor an alternative safety system.

The Cospas-Sarsat system, on the other hand, provides a unique and effective tracking and position-fixing system for 406MHz EPIRB's and is supported by a land-based infrastructure for initiating and controlling any necessary search and rescue operation on land or at sea.

The Cospas-Sarsat system provides complete global coverage using polar-orbiting satellites. EPIRB signals can be processed and lodged with a Local User Terminal within minutes of receipt if an LUT is in 'real-time' view.

Signals received from more remote locations are stored onboard the satellite and around three hours can pass before the message is passed on and rescue operations started.

The system will fix the position of a casualty to within a few miles. SAR forces can then be directed close enough to the scene to allow them to home-in on the EPIRB, using its secondary frequency of 121.5MHz.

The 406MHz EPIRB and the Cospas-Sarsat tracking system now form an integral part of the Global Maritime Distress and Safety System (GMDSS) — the subject of the next chapter.

Iridium and Odyssey will offer the mariner the first real alternative world-wide mobile satellite telephone services to Inmarsat — at the same time that Inmarsat is to offer its own 'Global Personal Communication System'. All three will provide cellular-compatible mobile terminals (hand-sets, etc), but through different types of satellite system.

In addition to the telephone/fax/data etc. communications provided by these new Mobile Satellite Systems, the mariner will want to be satisfied that adequate provision is being made for safety communications. Either over the MSS themselves or by alternative means.

What does seem clear is that these new systems will, finally, extend the ease of use of automatic cellular radio telephone systems to smaller craft, world-wide, at a truly affordable price sometime near the end of this century.

Meanwhile, you need to consider how to bridge the gap of the intervening years!

Chapter 9
The Global Maritime Distress and Safety System (GMDSS)

Introduction

Users of maritime radio and satellite communications equipment fall into two categories — those aboard 'SOLAS convention' vessels, which are obliged to fit particular equipment for distress and safety purposes (and are therefore known as 'compulsory-fitted') — and those aboard 'voluntary-fitted' craft, who can choose what equipment to fit from the range available and what to leave out.

Whether you are aboard a voluntary or a compulsory-fitted vessel, you will have access to the same GMDSS infrastructure both ashore and afloat, for distress and safety purposes, if you fit the same type of radio/satellite communications equipment.

The primary distress and safety communications facilities available during the twentieth century evolved from the original Morse telegraphy service; through to radiotelephone and radio telex services (on MF, HF and VHF) and now, additionally, into satellite communications.

Mariners relied mainly on an aural (listening) watch being maintained by radio operators onboard convention ships and in Coast Radio Stations. 'Automatic Alarm' devices offered a limited amount of automated alerting. Any rescue on the high seas was more likely to come from another vessel who had heard your distress call, than from any shore-based helper. That meant that once out of radio range of the coast, you were relying on another vessel being in radio range and picking-up your call for help.

The GMDSS, which is being introduced progressively over a seven-year period which started in 1992, is set to change that. Using a combination of marine radio and satellite systems from the preceding chapters, the GMDSS will change the emphasis of distress communication in two ways.

The first is to replace the aural watch with automatic alerting.

The second is to provide a shore-based infrastructure for the receipt of those automatic distress alerts (so that you no longer have to rely on another vessel being within radio range) and for the instigation and control of the required Search and Rescue (SAR) operation.

On-scene and bridge-to-bridge voice communications will continue to rely on CH16 VHF and 2182kHz voice.

The shore-based facilities provided within the GMDSS will vary from country to country, but will be chosen from the range available. Convention ships will be obliged to fit particular equipment to work with the facilities provided ashore, depending on their sea-area of operation.

Owners of voluntary-fitted craft need to understand how the GMDSS works and what facilities will be provided in their own area of operation, so they can choose the most appropriate radio/satellite communication equipment for their own needs.

The aim of this chapter is to explain the GMDSS — the shore based infrastructure and equipment used — so that you can make an informed choice from the communications facilities available.

General Concept of the GMDSS

The GMDSS uses two satellite constellations (Cospas-Sarsat and Inmarsat) and Digital Selective Calling (DSC) on VHF, MF and HF, as the primary means of distress alerting. If you want to be able to send an automatic distress alert through the GMDSS, you will need to fit alerting equipment which will work into one or more of the above facilities.

The 406MHz EPIRB distress alert works through the Cospas-Sarsat satellites (Chapter 8), to be passed on to a Local User Terminal (LUT) ashore for processing and transfer via a Mission Control Centre (MCC) and then to an associated Rescue Co-ordinator Centre (RCC). If the associated RCC is not the one best placed to handle the SAR operation (eg, if the casualty is off the coast of another country), the details of the casualty will be passed to another RCC — one which is better placed to instigate and control the SAR operation.

406MHz EPIRB's provide total global coverage with virtual 100% probability of detection, with position fixing to within a few miles. It can, however, take anything from 30 minutes to a few hours for the process to be completed and for a SAR operation to commence.

L-Band EPIRBs also provide satellite-based alerting — using the Inmarsat birds. There is a single dedicated CES for each of the four Inmarsat Ocean Regions. The CES will receive the alert and pass it on to an associated RCC for action. L-Band EPIRBs provide *instant* alerting and the alert itself contains the position of the casualty. The system does not employ doppler-analysis, as in the Cospas-Sarsat system. (See Chapter 8)

The position contained in the L-Band EPIRB is obtained from onboard navigation equipment (eg, a GPS unit), which can be integral to the EPIRB or separately connected. Whichever method is chosen, it is important that position information is kept up to date, as you may not get the opportunity for a final update when a disaster occurs.

SOLAS convention vessels *must* carry a satellite EPIRB (either the 406MHz or the L-Band type). These EPIRBs virtually guarantee your distress alert being received ashore, with minimal effort onboard, in a way that no other system can match.

Satellite alerting within the GMDSS can also be carried out by using the operational Satcom (A,B or C) equipment onboard. All three types are fitted with a 'distress button' — that red knob which will give you access direct to an RCC through a Coast Earth Station. Using normal Satcom equipment also allows you to customise your distress message if you have time to do so.

Most Inmarsat-M kit, although not 'GMDSS approved' and therefore not authorised as GMDSS equipment for convention ships, will allow the same 'panic button' access to an RCC ashore for voluntary-fitted vessels.

When you have passed a distress message direct to the RCC by using Satcom equipment, you should close down the channel and keep your equipment free to receive a call back from the RCC (unless the RCC has asked you to keep the channel open).

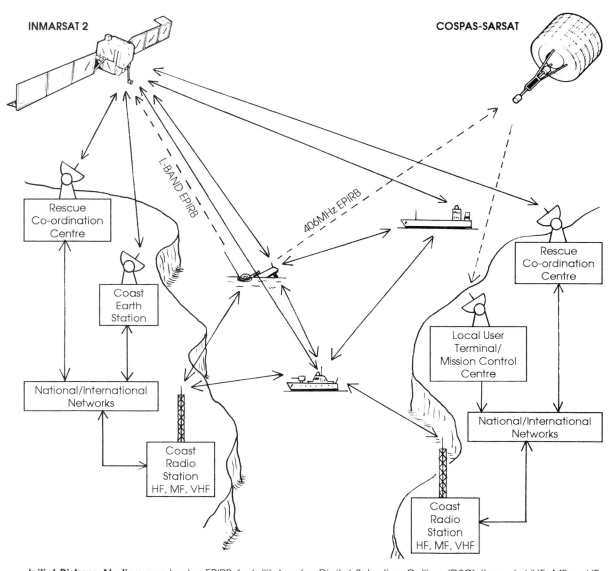

INMARSAT 2

COSPAS-SARSAT

L-BAND EPIRB

406MHz EPIRB

Rescue
Co-ordination
Centre

Coast
Earth
Station

National/International
Networks

Coast
Radio
Station
HF, MF, VHF

Rescue
Co-ordination
Centre

Local User
Terminal/
Mission Control
Centre

National/International
Networks

Coast
Radio
Station
HF, MF, VHF

Initial Distress Alerting can be by EPIRB (satellite) or by Digital Selective Calling (DSC) through VHF, MF or HF Coast Stations.

To-ship alerting will be by DSC and/or Inmarsat SafetyNET service, to Enhanced Group Call (EGC) receivers (mostly on 'convention' vessels, until the system becomes more widespread on voluntary-fitted vessels).

Two-way communications with the shore rescue authorities by Satcom, VHF, MF or HF — depending on equipment and position of casualty.

On-scene and SAR vessel communications by radiotelephone on VHF Ch16 and/or 2182kHz(MF). Radio-telex on 2174.5kHz may also be used.

Fig 9.1 General Concept of the GMDSS

How to Send a Distress Call
Inmarsat-C Terminal

Method 1
Sending a distress alert using Inmarsat-C SES terminal 'menus' (typical process).

1   Access the distress menus on your SES terminal.

2   Fill in the selections on the menus presented, firstly entering your
      vessel's name and position, and then manually entering as much other
      information as you can in the time available.
      (Some terminals will enter some information automatically.)

3   Specify the distress type as maritime and the nature of the distress
      from the list provided. (Sinking, on fire, abandoning, etc).

4   Select the nearest CES to your vessel within your ocean region.
      (This will normally also connect you with the closest RCC — the one
      associated with the CES. You may, however, select *any* CES within
      your ocean region, if that CES can connect you to an RCC which can
      help you, for example, by communicating in your native language.)

5   Confirm, by pressing the appropriate key(s), that you want to send the
      distress alert. (The SES will now automatically transmit your distress
      alert via the selected CES to its associated RCC.)

6   Wait for an acknowledgement from the CES. (If you do not receive
      an acknowledgement within 5 minutes, repeat the process.)

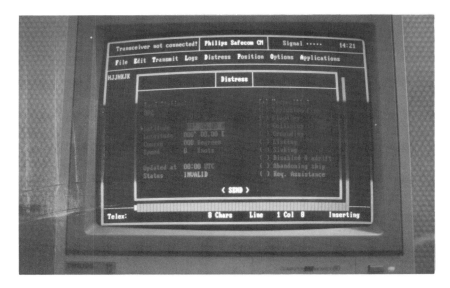

Inmarsat-C Distress Menu Screen. If time allows, you should always try to provide current information — eg of position and type of distress — before sending the message. Because of the large number of false alerts being received from Inmarsat-C terminals, the RCC will probably try to call you back to confirm your situation before mounting a Search and Rescue mission.

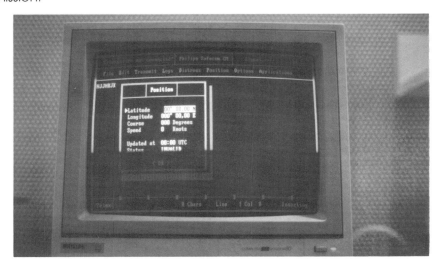

The position screen is associated with the distress screen and may be updated, at intervals, from your associated GPS. If the interval between updates is lengthy, you should consider a manual update in emergency situations.

How to Send a Distress Call
Inmarsat-C Terminal

Method 2

If you want to get an alert off instantly without going through the 'menu' process, you can send a distress alert using the remote distress button(s).

A typical process will require you to press the button(s), and hold it/them down for a specified period of time (typically 5 seconds). Note the following points about using this method of alerting.

Note 1

Sending a distress alert by pressing the distress button(s) sends only pre-programmed information dating from when it was previously entered (automatically or otherwise). Since then, your position, course and/or speed may have changed. If so, and assuming time permits, you should update the information sent to the rescue authorities by sending either an updated distress alert from your terminal or a more detailed distress priority message.

Note 2

Pressing the remote distress button(s) sends a distress alert immediately via the Inmarsat system to an RCC, whether or not your SES is engaged in message transfer, and whether or not your SES is logged-in to an ocean region.

Note 3

To avoid sending false distress alerts, do not press the remote distress alert button(s) except in a real emergency when you are in grave and imminent danger.

Note 4

Inmarsat-C is a store-and-forward message system. You cannot make direct, two-way contact with shore authorities in 'real time' using Inmarsat-C.

After you have sent the distress alert or distress priority message, set the automatic scan on your Inmarsat-C SES to scan only your current ocean region, to ensure that the RCC which received your message can return your call.

(The above information is adapted from the Inmarsat-C distress procedure provided courtesy of Inmarsat).

Some RCCs are fitted with their own Satcom terminal, so they may choose to use that terminal to re-establish a direct link with a casualty or to communicate with others who might be able to help.

The other method of distress alerting from sea is by Digital Selective Calling — either on VHF, MF or HF. DSC alerts can be picked-up by suitably equipped Coast Stations and also by convention vessels who are equipped for the GMDSS (as all should be by 1999). Using VHF/MF DSC, when compulsory-fitted ships are in range, could result in a quicker rescue than might otherwise be the case.

A number of Coast Stations (and some convention ships) maintain an HF DSC watch on designated frequencies (see FIG 9.2). Others are expected to join towards 1999.

Action by the 'First RCC'

Whichever shore-based system receives your alert — a Cospas-Sarsat LUT; an Inmarsat L-Band EPIRB CES; a DSC station on VHF, MF or HF — your message should result in a search and rescue operation being mounted which, hopefully should see you picked-up and safely brought ashore.

Communications with the Casualty

When a distress alert is received ashore, the controlling RCC will try to establish communications with the casualty. The method used will depend on how the alert was received.

If you send out a DSC alert on VHF Channel 70, in range of a VHF DSC Coast Station, the RCC will send a DSC acknowledgement on CH70 and will set watch on VHF CH16. You should set watch on CH16 yourself when your alert has been acknowledged and send a verbal Distress call and message. (Convention ships in range should also pick-up your CH70 alert and the Coast Station acknowledgement and will also set watch on CH16.)

DSC alerts on the MF frequency of 2187.5kHz, if picked-up by a shore-based DSC station, will result in the RCC acknowledging by DSC broadcast, also on 2187.5kHz, and setting watch on 2182kHz to await your Distress call and message.

If you use a Inmarsat-A, B, M or C Satcom terminal to send your alert and your initial distress message, the RCC will come back to you on a similar Satcom channel.

Where alerting is by EPIRB, the method of subsequent distress communications will depend on the position of the casualty and the shore-based infrastructure in the area concerned and any additional information you provide in your distress message.

Alerting Rescue Vessels

The RCC will alert Coastguard cutters, RNLI lifeboats, SAR aircraft etc using their own landline systems. Alerting potential rescue vessels from the multitude of craft at sea, going about their normal business (or pleasure), relies on a variety of broadcast methods in addition to the DSC broadcasts above.

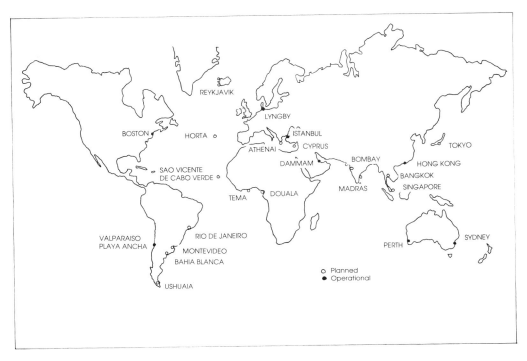

HF DSC Stations in the GMDSS

HF DSC Frequencies(kHz)
4207.5; 6312; 8414.5; 12577; 16804.5

(The total number of sites is expected to increase considerably as some countries identify additional sites and others, who have declared an interest in HF DSC but have not yet chosen a site, do so.)

Ships in distress can use one of two methods of alerting, using the listed HF DSC frequencies.

(i) The distress alert can be transmitted five times on the same frequency for each sending.

(ii) The alert can be sent once on each of five separate frequencies.

Suitably fitted convention vessels will normally employ a scanning-receiver to cover the HF channels, rather than retain a watch on one HF frequency only, or on all frequencies at the same time.

Fig 9.2 HF DSC Stations and Frequencies

NAVTEX

The NAVTEX system is used to send an initial distress alert when the casualty is within a NAVTEX station area of coverage. The type of NAVTEX receiver carried onboard convention ships will produce an audible alert on receipt of an initial distress message. The Master of the vessel will then set watch on CH16/2182kHz if in the area of the casualty.

SafetyNET

A message will be broadcast over the Inmarsat SafetyNET system to all convention ships who have Satcoms fitted as part of their GMDSS installation and who are operating within the general area of the casualty. The SafetyNET message will be picked-up by the onboard Enhanced Group Call receiver and, again, will sound an alarm onboard the vessel(s) concerned.

The SafetyNET system allows for 'Area Group Calls', whereby only those vessels within a particular area, rather than all vessels in the Ocean Region served by the satellite, are alerted. This is possible because EGC receivers hold details of the ship's position at all times and compare that position with the 'area group call' details in the SafetyNET message, and only accept those which are relevant.

Choice of Equipment

Sea-area of Operation

All convention ships are compelled to fit particular radio/Satcom equipment depending on their sea-area of operation. The sea-areas are designated as A1, A2, A3 and A4.

Area A1 will always be an inshore/coastal area. It can be anywhere in the world, as long as it is within radiotelephone range of a shore station providing continuous DSC coverage on VHF CH70. There will therefore be a large number of 'A1' areas but you could never go around the world without *leaving* area A1.

Sea-area A2 is represented by an area within radiotelephone range of a shore station which maintains continuous MF (2187.5kHz) DSC coverage (but excluding an A1 area, where the two overlap).

Sea-area A3 is represented by the ocean regions covered by the Inmarsat system but, where an area is already designated as A1 or A2 by virtue of VHF/MF DSC coverage, it will not also be considered as an A3 area.

Sea-area A4 is represented by the remaining areas of the world oceans — mainly towards the polar regions — where communication with the shore authorities is only possible by using HF radio.

It is the availability of 'continuous alerting' by one of the accepted GMDSS methods which determines which designator will apply to which area.

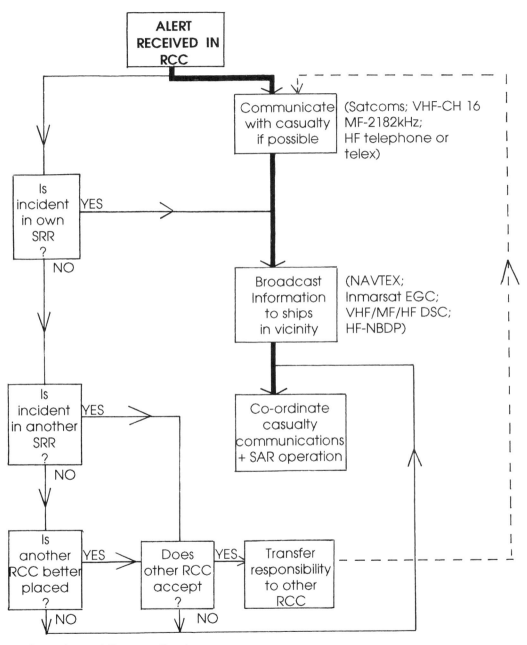

SRR = Search and Rescue Region

Fig 9.3 Action of First RCC on Receipt of a Distress Alert

Functionality

For each type of sea-area (A1, 2, 3 or 4) which a 'convention' vessel is expected to operate in, the vessel concerned must carry radio communication equipment which will allow it to carry out the following nine functions.

1   Sending ship-to-shore distress alerts.
2   Receiving shore-to-ship distress alerts.
3   Sending and receiving ship-to-ship distress alerts.
4   Sending and receiving SAR co-ordination messages.
5   Sending and receiving on-scene communications.
6   Sending and receiving 'locating' signals.
7   Receiving and sending Maritime Safety Information (MSI).
8   Exchanging 'general radio communications' with shore-based radio systems or networks.
9   Sending and receiving bridge-to-bridge communications.

The methods that will be used by convention ships to meet those requirements of functionability are summarised in Fig 9.4 — Carriage Requirements. A brief study of the table will give voluntary users some idea of a sensible fit for their own needs.

You can see that no convention ship is allowed to go to sea without five particular items of radio equipment — a VHF R/T (with DSC); NAVTEX; an EPIRB; a 9GHz Search and Rescue Transponder; and a hand-portable VHF R/T unit. (In fact, they must carry two each of SART and portable VHF).

If the vessel is to go out of range of coastal VHF-DSC stations, then the EPIRB must be a satellite EPIRB (406MHz or L-Band).

If venturing beyond the range of shore-based VHF DSC stations, additional radio/Satcom equipment must be fitted.

Although MR R/T is listed as an option for Area A2, there is a restricted choice of MF-only equipment available — most manufacturers now producing combined MF/HF transceivers. That is certainly the case for voluntary-fitted vessels, who can therefore consider areas A2/A3 as being the same.

Looking down the A3 area, you have a choice in two main spheres. The first is with your choice of EPIRB. Do you go for the 406MHz Cospas-Sarsat item or the Inmarsat L-Band EPIRB? At the time of writing there was a reasonable range of 406MHz EPIRB's but few L-Band units. There was also a considerable range of prices. Either (or both) of those factors might affect your own choice.

The next factor to consider is the area you wish to operate in, remembering that the Inmarsat system does not have complete global coverage, but that the (polar) areas not covered might be areas to which you will never venture.

The final factor in choice of EPIRB is whether you prefer the Cospas-Sarsat method of position fixing — to within a few miles in all cases, without manual input to the unit combined with an alert period of up to a few hours — or the 'instant' Inmarsat alert which may contain a position (from your GPS) which is a few hours old.

		AREA OF OPERATION			
		A1	**A2**	**A3**	**A4**
	EQUIPMENT				
1	VHF R/T with DSC	✔	✔	✔	✔
2	MF R/T with DSC	—	✔	✔ or item 3	—
3	MF/HF R/T with DSC and NBDP	—	—	✔ or item 5	✔
4	NAVTEX	—	✔	✔	✔
5	Inmarsat A, B or C SES; plus EGC Receiver	—	—	✔ or item 3	—
6	406 MHz Satellite EPIRB	✔ or item 7/8	✔ or item 7	✔ or item 7	✔
7	L-Band Sat EPIRB	✔ or item 6/8	✔ or item 6	✔ or item 6	—
8	VHF EPIRB	✔	—	—	—
9	9GHZ SART	✔	✔	✔	✔
10	Portable VHF Tranceiver	✔	✔	✔	✔

The table shows the type of radio/Satcom equipment to be carried by compulsory-fitted vessels — but not the number of units. The information can be used as a guide by voluntary-fitted craft to the type of equipment appropriate to their own area of operation.

FIG 9.4 GMDSS Carriage Requirements (Type of Radio Installation)

Long-range Communications

The other main area where convention vessels have a choice is in the provision of facilities for long-range (ie — other than VHF) communications. The choice is firstly between terrestrial radio (MF/HF) and Satcom equipment and secondly, where Satcom kit is chosen, between a unit which provides voice communications (Inmarsat-A/B) and one which provides text messaging only (Inmarsat-C).

The voluntary-fitted vessel who wants to fit equipment which will work into the GMDSS infrastructure can fit less expensive (voice) kit than compulsory-fitted ships. The choice for smaller leisure craft is between a Marine SSB or an Inmarsat-M (voice) or an Inmarsat-C (messaging) Satcom unit. Price will definitely be a factor for most leisure sailors and fishermen. There is a considerable price difference between Marine SSB suitable for voluntary-fitted craft and most Satcom equipment.

Like other things in life however, we can expect to see some change during the latter half of the 90's decade, as the availability, size and cost of Satcom equipment gradually comes down and more and more radome-covered antenna units appear in our marinas.

The 9GHz SART — compulsory fit for convention vessels, but 'take it or leave it' for the rest of us — is the item which bridges the gap between the position accuracy of the Cospas-Sarsat system and the casualty in the water.

The SART will put out a signal which will appear as a distinctive trace on the radar screen of an approaching vessel or SAR aircraft. The SART is activated by the pulse received from the radar of an approaching ship/aircraft and will emit a signal which can be detected some miles away (not less than 10 miles for an approaching craft with an antenna height of 15M is quoted). As the Cospas-Sarsat system should bring your rescuer to within a few miles of your position the SART should, theoretically, guide them across the remaining distance.

There are units available which combine a 406MHz EPIRB with a SART and which might be considered more practical than separate units when limited space is a factor.

The final item on the GMDSS list is the hand-portable VHF transceiver. A hand-held unit can come in useful when your main antenna has been brought down or when you have lost primary VHF communications for some other reason. Also in the liferaft!

Although limited in range at sea level, the hand-held unit will allow you to communicate some considerable distance when SAR aircraft are involved, as such aircraft are also fitted with CH16 VHF equipment. The hand-held unit is also practical in the smallest of life-rafts — but do not try to communicate through wet canvas if you can avoid it. The low power output from a hand-held set really needs a clear transmission path.

Prevention — Better than Cure

Also included in the list were three items which should help you avoid trouble, at least from outside forces such as the weather, or navigational hazards.

There are radiotelephone broadcasts of weather, Nav warnings etc, which many mariners will take advantage of. It is not always easy to monitor all broadcasts personally however, and various methods of automatically providing 'Marine Safety Information' (MSI) have been provided within the GMDSS.

The three main sources of MSI for convention vessels (and which can be monitored by anyone else with the appropriate kit) are NAVTEX, Inmarsat EGC, and HF NBDP (telex broadcast).

Convention ships, as before, must fit equipment to receive MSI appropriate to their area of operation. This does not necessarily follow the A1/A2/A3/A4 pattern.

If a vessel is sailing in an area with NAVTEX coverage, then it must fit NAVTEX. This is the primary means of broadcasting MSI for areas up to 200 miles from the coast where NAVTEX is provided ashore.

Ships going outside the area served by NAVTEX must fit either Inmarsat EGC or HF NBDP, depending on the facilities provided from shore.

Unfortunately, not all countries who subscribe to the GMDSS principle and on whom mariners depend for the broadcast of safety information, have implemented arrangements for comprehensive MSI broadcasts on Inmarsat EGC/HF-NBDP. Mariners meanwhile need to remain aware of all available sources of information and take advantage of those which give best coverage of their own area of operation.

Another drawback on the Inmarsat EGC front, for voluntary-fitted vessels, is a lack of stand-alone EGC receivers. EGC is an integral part of an Inmarsat-C terminal. As most convention ships will include at least one Inmarsat-C terminal onboard, there is little apparent need for a separate EGC receiver.

Manufacturers are also providing an EGC 'add-on' for Inmarsat-A which, again, reduces any need for a stand-alone unit (bringing, as it would, a need for yet another antenna).

What we are likely to see however, is a touch of entrepreneurship from those people who serve the radio amateur/leisure craft market — those who produced 'small ships' NAVTEX units, or who write software to emulate NAVTEX, NBDP etc to work on your laptop PC.

Surely one or more of these irrepressible innovators will give us what we need before too long?

NBDP (SITOR) Broadcasts
It seems likely that those countries who are providing a HF DSC watch will also provide a NBDP broadcast facility — if there is a need for an HF DSC watch then there is an implicit need for complimentary MSI broadcast facilities.

Hopefully, many of the stations which currently provide safety information by radio-telex/radiotelephone broadcast will continue to do so — to help those who do not require a full GMDSS fitting to keep out of trouble.

Areas A1 and A2
In addition to the A4 stations listed in Fig 9.2; Fig 9.5 lists those countries which have given intention to declare A1 or A2 areas, and to provide the appropriate (VHF DSC/MF DSC) facilities ashore.

	A1(VHF)	A2(MF)	A3/A4(HF)
Argentina	N	Y	Y
Australia	N	N	Y
Belgium	Y	Y	N
Brazil	Y	N	Y
Cameroon	Y	N	Y
Canada	?	?	?
Cape Verde	Y	Y	Y
Chile	Y	Y	Y
China	Y	Y	Y
Croatia	Y	?	N
Cyprus	Y	Y	Y
Denmark	Y	Y	Y
Egypt	N	Y	N
Estonia	?	?	N
Finland	Y	Y	N
France	Y	Y	?
Germany	Y	Y	N
Ghana	Y	Y	Y
Greece	Y	Y	Y
Hong Kong	Y	N	Y
Iceland	?	Y	Y
India	N	N	Y
Ireland	N	Y	N
Italy	Y	Y	Y
Japan	N	Y	Y
Latvia	?	?	?
Lithuania	?	?	?
Netherlands	Y	Y	N
N Zealand	N	N	Y
Norway	Y	Y	N
Portugal	Y	Y	Y
Rep of Korea	?	?	Y
Saudi Arabia	?	Y	Y
Singapore	Y	Y	Y
Spain	?	?	Y
Sweden	?	Y	N
Thailand	Y	Y	Y
Turkey	Y	Y	Y
UAE	Y	?	Y
UK	N	Y	N
USA	Y	?	Y
Uruguay	Y	Y	Y

Key: Y = Yes; N = No; ? = Information not available

FIG 9.5 Countries Providing/Intending to Provide A1(VHF DSC); A2(MF DSC); A3/A4 (HF DSC) Facilities

Summary

The GMDSS, which should be fully implemented by 1999, provides more scope for automatically alerting shore authorities of problems at sea than previous systems.

The type of equipment which compulsory-fitted vessels must carry and the facilities provided by the participating countries around the world can be used as a guide by owners of voluntary-fitted craft when choosing what to fit themselves, so that they have the best chance of being rescued in the event of the ultimate disaster at sea.

The voluntary-fit can utilise lower-cost equipment than the compulsory-fitted vessel to gain access to similar facilities ashore (eg Inmarsat-M Satcom *vice* Inmarsat-A/B) and also for the reception of safety broadcasts (NAVTEX, NBDP etc).

Some systems which might appear to be the best choice for the future may not be operational in the area you require now, thus forcing an interim choice (which might have to be changed later), or causing you to choose to do without meanwhile.

Some items will remain a more affordable part of the GMDSS infrastructure for the foreseeable future. These include your Marine VHF (with or without DSC) and Satellite EPIRB's (with or without a SART). Others may be out of (financial) reach.

The big choice for all mariners is between terrestrial radio (MF/HF) and Satcom equipment for longer range communications. If that choice is not made for you (eg, because you intend a north-about trip across Canada!), then other factors will have to be compared.

Choosing the best fit for your own circumstances is the subject of the next chapter.

Chapter 10
What's Right for You

Introduction

The mariner, as we can see from the preceding chapters, is faced with a considerable choice of radio communication systems — both terrestrial (VHF, MF, HF) and satellite based — and different types of installation which can be used to access any one facility.

Whilst many vessels, being 'compulsory-fitted', are told what marine systems they must cater for (and the quantity and quality of equipment they need to fit), the vast majority of sea-going craft are 'voluntary-fitted' and, as such, can choose whether or not to fit radio at all. Should they decide to fit any particular type of radio communication system, owners of voluntary-fitted craft can also choose from a much wider range of marine equipment which will carry out the functions required. Or they can choose from alternative, non-marine methods, if they feel that such alternative services are more appropriate to their own needs and circumstances.

Earlier chapters attempted to explain the various choices for each individual category of system — short-range/long-range telephone, etc. In this chapter we take an overall view of your likely communication requirements — short and long term — in an attempt to make the choice of systems and methods one which you will be happy with.

Questions to be Asked

The questions you will need to ask yourself can be put into one of four categories, namely
(i) Assess your communication requirements.
(ii) What methods are available to meet your requirements?
(iii) What constraints are you up against?
(iv) What will be the impact on other onboard systems?

Assessing Your Requirements

The first thing to consider is the area you will be cruising or operating in and including any area you will be taking passage through to get there. You need to know what radio communication facilities you can access from those areas concerned and then you need to decide which of those facilities you want to take advantage of.

When considering the above, ask yourself which items are essential and which are desirable but not essential.

Meeting Your Requirement

You then need to establish what marine radio equipment is available that will allow you to access the facilities you have earmarked, both essential and desirable.

You can also identify any (non-marine) alternative and weigh the benefits/drawbacks of each against the other(s).

Constraints

The first constraint would be that of the compulsory-fit. If you come into this category you will be *told* the type of equipment you have to install — with limited choice. If you fall into that category then you might as well start from there and identify only the items which might be missing.

For the majority of owners, those who are looking for equipment to fit voluntarily, the regulatory constraints are somewhat different and leave much more choice to the individual.

For USA craft; although you may not be compelled to fit VHF, you will not get a licence to fit SSB or Satcom equipment unless you fit a VHF set first. I have to admit to a liking for this particular regulation. The versatility of Marine VHF equipment, considering its simplicity of installation and operation — when combined with its low cost — make it a virtually indispensable piece of marine safety equipment. There is no alternative for direct Port/Harbour/Pilot communications and it offers free access to local weather and other safety information to a degree not provided by any non-marine system. It is difficult to see how anyone would consider fitting more expensive long-range radio/Satcom equipment without first fitting VHF. But that option does remain open in most countries.

The table in Fig 10.1 is provided to help you select radio equipment which will meet your desires/needs.

First, having read all the previous chapters, move along row 'a' and enter a 'C' where you *currently* want a particular facility and 'F' where you perceive a *future* need.

Next, you need to establish what facilities are available (or likely to become available) in the area(s) in which you intend to operate. This can be done by working through the considerable range of publications available from government communication agencies; the Coastguard; yachting and fishing almanacs; and amateur radio publications.

Your marine electronics supplier should hold a range of maritime publications and your local ham radio shop will hold those directed at the amateur market. You should not expect to have to *buy* them all — if the supplier wants to sell you his equipment, get him to prove that you will be able to access the type of service you require, once fitted.

Armed with the above information, work along row 'b' and enter C (current) or F (future) to indicate the availability of a particular service. Also note how a particular service is provided (eg. short-range distress alerting by VHF DSC; long-range by HF DSC/Satcom).

The next step, row 'c', is to decide what facilities are *essential* as opposed to those which are merely desirable. This is a personal decision, based on what you want, what is available (and how) and what is not.

If you have an essential requirement which does not appear to be available from your first review of facilities, then consider how it can be made available to you: eg — if there are no short-range (VHF radio) distress alerting facilities in the area, how will you kit yourself out to send an alert from those waters?

When you have listed your essential and desirable requirements, you can work through the remaining rows and compare the various ways of meeting your needs. You will notice

	Distress Alert		Distress Traffic		Inter-ship		Ports Pilots & VTS	
	Short-range (VHF)	Long-range	Short-range (1)	Long-range (1)	Short-range	Long-range		
Desirable (C/F)								a
Available (C/F)								b
Essential (C/F)								c
Provided by Marine VHF TRX								
- USA Chan (3)	(2)		✓		✓		✓	d
- Int'l Chan(4)	(2)		✓		✓		✓	e
Marine SSB TRX	(2)	(2)	✓	✓	✓	✓	(10)	f
Inmarsat								
- A/B		✓		✓	(8)	(8)	(10)	g
- C (9)		✓		✓	(8)	(8)	(10)	h
- EGC		(11)						i
- M		✓		✓	(8)	(8)	(10)	j
Future Mobile Satellite Systems								
-P (13)								k
IRIDIUM (13)								l
Odyssey (13)								m
EPIRB								
- 406MHz		✓						n
- L-Band		✓						o
- 121.5/243		✓						p
NAVTEX RX		(11)						q
Weatherfax RX								r
Cellular Radio							(10)	s
Amateur S/W TRX			(7)	(7)	(14)	(14)		t
Broadcast RX			(7)	(7)	(7)	(7)		u

Notes

(1) Two-way exchange of distress traffic.

(2) Normally voice but automatic alerting possible with DSC option.

(3) USA channels needed for access to continuous WX channels.(Visitors please note.)

(4) Recommended for USA/Canadian vessels going abroad.

(5) At specified broadcast times from available stations.
  Add-on units available for f, s and t for NAVTEX/SITOR/Fax broadcasts.

(6) With add-on unit.

(7) Receive-only mode.

(8) A pay-service via satellite channels.

(9) Inmarsat-C is a store-and- forward messaging system (not voice).

Fig 10.1 Facilities Assessment and Equipment Selection Chart

PC Telephone Short-range	PC Telephone Long-range	PC Telex & Data	Weather Forecast Local & Coastal (5)	Weather Forecast High Seas (5)	Nav Warnings Local & Coastal (5)	Nav Warnings High Seas (5)	News Ent'ment B'casts	Amateur Radio Service	
									a
									b
									c
✓			✓		✓				d
✓			✓		✓				e
	✓	(6)	✓	✓	✓	✓	✓	(7)	f
	✓	✓	(8)	(8)	(8)	(8)	(8)		g
		✓	(8)	(8)	(8)	(8)	(8)		h
			✓		✓				i
	✓	(12)	(8)	(8)	(8)	(8)			j
✓	✓								k
✓	✓								l
✓	✓								m
									n
									o
									p
			✓		✓				q
				✓					r
✓									s
(15)	(15)		✓	✓	✓	✓	✓	✓	t
(7)	(7)		✓	✓	✓	✓	✓	(7)	u

(10) PC Telephone/Telex calls can be made to port/harbour master ashore.
(11) Shore-to-ship only.
(12) Limited data facility.
(13) Proposed global personal telephone service - full range of facilities to be determined.
(14) Using frequencies in the Amateur Service
(15) Marine Channels - Receive only - non-commercial 'Phone-patch' facilities in the Amateur Service.

Final Reminder:
When you choose one type of radio equipment against another, you are not only choosing the facilities you *will* get - you are also choosing *to do without others*.

that for some services there is only one type of equipment which will provide direct, full function access to a particular facility — eg only a Marine VHF set will allow you to make two-way radio contact with Port Authorities on their marine frequencies.

For others (eg NAVTEX and Weatherfax broadcasts), the transmissions can be copied by using dedicated equipment or by use of an add-on facility to another type of radio set (eg Marine SSB).

The thing to remember is that where a particular piece of equipment is being shared between facilities — whether it is the radio itself that is being shared or the laptop computer with the NAVTEX/SITOR/Weatherfax software — you cannot use either shared facility to do more than one thing at any one time. If you have a requirement for a virtually continuous NAVTEX/Weatherfax or other type of watch, you will need to consider having equipment dedicated to that facility rather than shared with some other need.

Constraints — Physical

There are some physical constraints which need to be taken into account when deciding which equipment to buy and fit. That is why you need to look ahead at your likely total future requirement for radio (and other) equipment before making a choice.

Is there enough room below for all the kit you want to fit without compromising the space required for other facilities? (The galley is *not* a suitable place for overflow radio/electronics equipment.)

How many antennas are going to be required now and in the future? What type of antenna (wire/whip) is suitable for your desired equipment? How can you best use the space available up-top to cater for all your needs? *Is* there enough space for all or will compromises have to be made?

Constraints — Type-Approval

All marine radio equipment which has a facility to transmit must be 'type-approved' by the country of registration of the vessel concerned. That means that you cannot legally fit a Marine VHF onto a US registered yacht unless the VHF carries an FCC type-approval number. A UK craft can only fit equipment which is on the UK type-approval list.

This can be galling for a UK sailor visiting the USA and seeing an alternative, full-function radio set at a much lower cost to similar products on sale in the UK, but which cannot legally be fitted to the UK vessel because the equipment is not on the UK type-approval list. It does not necessarily mean that the equipment is not up to some superior UK standard, it may only mean that the manufacturer has not applied for UK type-approval.

Items like broadcast receivers, NAVTEX/SITOR/Weatherfax decoders for fitting to your Marine SSB/short-wave ham set and other receive-only/add-on equipment are not normally subject to the same type-approval constraints as equipment with a transmit facility.

Constraints — Power Consumption

Whenever you are considering adding to your radio/electronics equipment provision, you need to review the total power consumption requirements of all electrical onboard equipment against the abilities of your available power supply.

Radio equipment has two levels of power drain — a lower level in the receive/idle mode and a higher level in the transmit mode. You may consider that the transmit mode on a Marine SSB set will be something you rarely use, perhaps only for the occasional brief telephone call, and that it will therefore not impact too much on your power source. But what about the emergency situation, when you suddenly have a need to transmit not only one brief call, but to transmit on-and-off over a period of hours, or even days — perhaps when your engine (and with it any ability to recharge your batteries) is disabled?

A robust power supply, one which can easily meet the needs of all your onboard equipment (present and planned), is essential if you are to get value from your radio installation. And don't forget that radio equipment will only quote the power drain for the radio itself. If you are buying a set-up with add-on units — automatic R/T, laptop PC and printer etc — you need to tote-up the power requirements of those peripherals in addition to those of the radio set itself.

Constraints — Mutual Interference

All radio and other electronic equipment, including the peripheral equipment mentioned in the previous paragraph, can be the source of interference for other equipment and can itself be interfered with.

The equipment itself must be correctly installed and may have to be shielded from other equipment or installed with enough separation to reduce interference to acceptable levels.

Antennas also need to be physically separate vertically and/or horizontally if the output from a transmit antenna is not to be fed straight into the receive unit of another piece of equipment, especially if they are designed for use on similar frequencies.

There are also 'radiation-safe distances' for some types of antenna. The microwave dish enclosed in a Satcom radome pushes out a concentrated microwave signal in the direction of the satellite and also has strong 'sidelobes' which can be harmful to humans (and other animals) if over-exposed at too close a range. If a satellite antenna is to be fitted onboard at a low level — in a position where anyone aboard can come close to the radome — make sure you (and others aboard) understand the need to minimise exposure to microwave radiation at close distance and for prolonged periods (see manufacturers literature for individual terminals for exclusion zones).

The Satcom antenna should not be obstructed or 'shadowed' by superstructure, people, terrain or buildings, or it will not transmit/receive effectively. The antenna requires a clear path towards the satellite if it is to work properly. You can do something about correct siting onboard and, by recognising the need for a clear path, you may be able to manoeuvre into a more favourable position in enclosed waters if communications are affected by hills or buildings.

Constraints — Secrecy

It is quite normal to be able to overhear conversations on ship-to-shore radiotelephone channels between other mariners and those they have called ashore, especially when you have to select a channel and await an opportunity to call the Coast Radio Station or have been asked by the CRS to stand-by on an occupied channel. It is neither illegal nor improper to overhear the conversation of other people in these circumstances.

What you cannot do is to disclose any details of such conversations — neither the fact that 'Albert on the *Highland Roundhead* was speaking to Jean in the office', nor any information about the content of any call you may have overhead (ie — what Albert and Jean were actually talking about).

Neither can you (legally) take advantage of any information you may have overhead. People will, of course. This is why commercial fisherman go to such trouble to disguise the true details of their catch when talking on an 'open' channel and why many of those same fishermen have taken to the 'secrecy' facility on Autolink RT!

One thing you will learn if you choose to go down the amateur radio route is that many of the radio/telephone facilities you previously thought were 'secret' can be intercepted by suitable equipment. It is normally the case that the onus would be on any offended party to prove that another had acted illegally and so, if you maintain your silence on any (accidentally or deliberately) overhead conversation, then you are not likely to fall foul of the authorities on that particular score.

Constraints — Who's Boat is it?

Radio equipment can only be installed and operated onboard any vessel 'on the authority of the Master'. This means that you cannot take your mobile ham radio rig onto someone else's boat and operate it without their permission. 'It's all right, I'm a licensed radio ham' is not enough. If the owner/master does not want your rig onboard, then you cannot legally install it. If you intend taking a ham rig onto a chartered yacht, then clear it with the charter company well in advance and in writing.

When ham radio rigs are installed on any vessel, either your own or someone else's (with their permission), then the electrical installation of the equipment must comply with the rules of the country of registration of the vessel concerned, in terms of safety of installation and electrical interference etc.

Constraints — Ship Station (Radio) Licence/Amateur Licence

In addition to the 'type-approval' requirement, you need a Ship Station (Radio) Licence to cover all transmitting radio equipment (including radar) fitted onboard any vessel. The basic licence of most countries will cover VHF, radar and EPIRBs, with a facility to include Marine SSB and/or Satcom equipment. Cellular radio systems do not normally require a ship's radio licence, nor does receive-only equipment.

Radio hams will hold a radio licence to install and operate equipment from their normal location (eg, their own house) and the licence also allows the operation of a mobile station (eg, from a road vehicle, a sea-going vessel or river craft). The ship's radio licence does not normally need to be altered to include the temporary installation of a maritime mobile station but, if in doubt, ask your licensing Authority (listed in Appendix Q).

Constraints — Operators Qualifications/Authority to Operate.

In addition to the Ship Station Licence (for equipment), the operator also needs to be licensed. Some countries (eg, the UK) require individuals to pass an examination before being allowed to operate any ship's radio equipment. Separate examinations and qualifications are in place for VHF and for Marine SSB equipment. Ship's officers and crew will have to pass examinations to show that they are competent to operate the radio/Satcom/DSC equipment in a GMDSS installation.

Owners of voluntary-fitted vessels in the USA do not have to pass an examination to be allowed to operate Marine VHF equipment on domestic voyages. If they want to communicate on VHF with a foreign station however, or if a Marine SSB set is to be used, even with stations in the USA, then the Restricted Radio Operators Permit must be obtained. Different rules for different countries. If in doubt — ask!

Constraints — Ports and Harbours

There are constraints on the use of radio equipment for Public Correspondence/Inter-ship communications when alongside in ports or at anchor in various harbours around the world. The constraints can range from a complete ban on the use of radio (transmitting) equipment, to allowing the use of equipment to communicate 'with the nearest coast station'.

Check your almanac, ALRS or other publication for any restrictions in the particular port you wish to visit or, if in doubt, refrain from transmitting when in harbour.

In harbours near military installations especially in foreign ports, it is prudent to leave your radio equipment switched off when alongside to avoid any misunderstandings if you are visited by the port authorities.

Summary

Compulsory-fitted vessels are told what equipment to fit — with limited choice — depending on where they will operate. Owners of voluntary-fitted craft (the vast majority) can use the guide-lines for compulsory-fitted vessels to decide what facilities they themselves want to take advantage of, and can then choose from a much wider range of equipment when putting together their preferred installation.

There are a number of constraints which are common to all, including licensing, operators qualifications and the need for secrecy. Other constraints, like the availability of space for the antennas required for a number of systems, differ from craft to craft.

As we anticipate a build-up of radio (and other electronic equipment) we need to ensure that the onboard power supply is capable of meeting the total requirement in the worst imaginable circumstances.

But we are entitled to remember that, provided we remain within the laws and regulations of our country of registration and those of any other country we visit, and that we act in a responsible manner — then we can choose whatever type of equipment we want to fit and select which services we care to utilize.

When you have made that choice and when you take for granted that which Marconi put so much effort into proving — namely, that radio waves can stretch across the Atlantic Ocean — then you should enjoy your investment in electromagnetic energy and use it to go even further than Mr M himself could achieve. If that is possible.

THE LANGUAGE OF MARITIME MOBILE AND AMATEUR RADIO, THE PHONETIC ALPHABET AND MORSE CODE SYMBOLS

	Word	Spoken As	Morse Character		Word	Spoken As	Morse Character
A	Alfa	**AL**-fah	• —	B	Bravo	**BRAH**-voh	— • • •
C	Charlie	**CHAR**-lee	— • — •	D	Delta	**DELL**-tah	— • •
E	Echo	**ECK**-oh	•	F	Foxtrot	**FOKS**-trot	• • — •
G	Golf	**GOLF**	— — •	H	Hotel	hoh-**TELL**	• • • •
I	India	**IN**-deeah	• •	J	Juliett	**JEW**-lee-**ETT**	• — — —
K	Kilo	**KEY**-loh	— • —	L	Lima	**LEE**-mah	• — • •
M	Mike	**MIKE**	— —	N	November	No-**VEM**-ber	— •
O	Oscar	**OSS**-car	— — —	P	Papa	pah-**PAH**	• — — •
Q	Quebec	keh-**BECK**	— — • —	R	Romeo	**ROW**-me-oh	• — •
S	Sierra	see-**AIR**-rah	• • •	T	Tango	**TANG**-go	—
U	Uniform	**YOU**-nee-form	• • —	V	Victor	**VIK**-tah	• • • —
W	Whiskey	**WISS**-key	• — —	X	X-Ray	**ECKS**-ray	— • • —
Y	Yankee	**YANG**-key	— • — —	Z	Zulu	**ZOO**-loo	— — • •

The syllables to be emphasised are shown in capital letters

No.	Codewords	Spoken As	Morse		No.	Codewords	Spoken As	Morse
0	NADAZERO	NAH-DAH-ZAY-ROH	— — — — —		1	UNAONE	OO-NAH-WUN	• — — — —
2	BISSOTWO	BEES-SOH-TOO	• • — — —		3	TERRATHREE	TAY-RAH-TREE	• • • — —
4	KARTEFOUR	KAR-TAY-FOWER	• • • • —		5	PANTAFIVE	PAN-TAH-FIVE	• • • • •
6	SOXISIX	SOK-SEE-SIX	— • • • •		7	SETTESEVEN	SAY-TAY-SEVEN	— — • • •
8	OKTOEIGHT	OK-TOH-AIT	— — — • •		9	NOVENINE	NO-VAY-NINER	— — — — •

All syllables have equal emphasis

Punctuation marks and other symbols are made-up by running two letters together, eg.

Mark	Written As	Morse Symbol	From eg.	Mark	Written As	Morse Symbol	From eg.
Full Stop	.	• — • — • —	AAA	Comma	,	— — • • — —	GW
Colon	:	— — — • • •	OS	Question Mark	?	• • — — • •	UG
Hyphen/Dash	-	— • • • • —	BA	Slant	/	— • • — •	XE
Open Brackets	(	— • — — •	KN	Close Brackets	)	— • — — • —	KK
Quotation Mark	"	• — • • — •	RR	Error	(not written)	• • • • • • • •	(8 x e's)

173

HOW TO LEARN MORSE CODE WITH MINIMUM PAIN.

(1) Make-up small cards – one for each letter/figure/mark – with the Morse character on one side and the symbol to be learned on the other, eg:

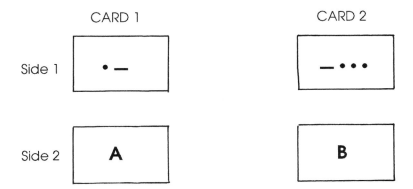

(2) Learn A, B and C on day one, and three new, additional characters on each day following – always going over all you have learned to date, eg.

Day 1 – A, B, C; Day 2 – D, E, F & A, B, C; Day 3 – G, H, I & A, B, C, D, E, F etc.

Do this by using the first part of the day on the new symbols only, and the remainder of the day on all symbols.

(3) Use "dit" and "dah" rather than "dot" and "dash" – and drop the "t" for "dit", except where it is the last digit on a character, eg.
di-dah (a); dah-di-di-dit (b).

(4)   When you have learned all the characters, borrow or buy some audio tapes, with slow-speed Morse pre-recorded.

(5)   In addition to the basic characters shown here, there are a large number of pro-words and abbreviations – you will need to know these if you want to get into operating, or listening on amateur service Morse bands – but they are not required for listening to, eg, Morse weather broadcasts.

(6)   The above process can also be used for learning the phonetic alphabet.

INTERNATIONAL CALL-SIGN ALLOCATIONS

All call-signs are derived from the ITU "International Allocation of Call-signs" table, reproduced below. Call-signs for particular types of "station" follow a pattern, as described.

(a)   The first character, or first two characters, will identify the nationality of the station.

(b) The formation of the call-sign will follow particular principles, depending on the type of station, eg.

(i) Coast Station – Maritime Mobile Service: Call-sign consist of three characters, where the first one or two characters can be either a letter of a digit, and the third character will be a letter. Alternately, a call-sign may have up to three additional characters, all of which would be digits eg.

> GKA = UK = Portishead Radio
> KZA917 = USA = Convent Marine Operator

(ii)   Ship Stations: Call-signs can be

– 2 characters and 2 letters; or
– 2 characters, and 2 letters and one digit, or
– 1 character, and one letter and four digits', or
– 2 characters, and one letter and four digits.

(iii) Amateur Service: Call-signs consist of:

Prefix = one/two characters identifying the nationality of the station;
followed by: a one-digit number
followed by: a suffix of up to 3 letters.

Amateur call-signs often additionally identify a more precise geographical location within a nation (eg. Prefix GM = Scotland, within the United Kingdom; KP2 = US Virgin Islands). Similarly, Maritime Service call-signs may distinguish a related nation (eg. a protectorate) of a country.

Call Sign	Allocated to	Call Sign	Allocated to
AAA-ALZ	United States of America	EIA-EJZ	Ireland
AMA-AOZ	Spain	EKA-EKZ	Russian Federation
APA-ASZ	Pakistan	ELA-ELZ	Liberia
ATA-AWZ	India	EMA-EOZ	Russian Federation
AXA-AXZ	Australia	EPA-EQZ	Iran
AYA-AZZ	Argentine Republic	ERA-ESZ	Russian Federation
A2A-A2Z	Botswana	ETA-ETZ	Ethiopia
A3A-A3Z	Tonga	EUA-EWZ	Belarus
A4A-A4Z	Oman	EXA-EZZ	Russian Federation
A5A-A5Z	Bhutan	E2A-E2Z	Thailand
A6A-A6Z	United Arab Emirates	FAA-FZZ	France
A7A-A7Z	Qatar	GAA-GZZ	United Kingdom
A8A-A8Z	Liberia	HAA-HAZ	Hungary
A9A-A9Z	Bahrain	HBA-HBZ	Switzerland
BAA-BZZ	China	HCA-HDZ	Ecuador
CAA-CEZ	Chile	HEA-HEZ	Switzerland
CFA-CKZ	Canada	HFA-HFZ	Poland
CLA-CMZ	Cuba	HGA-HGZ	Hungary
CNA-CNZ	Morocco	HHA-HHZ	Haiti
CQA-COZ	Cuba	HIA-HIZ	Dominican Republic
CPA-CPZ	Bolivia	HJA-HKZ	Colombia
COA-CUZ	Portugal	HLA-HLZ	South Korea
CVA-CXZ	Uruguay	HMA-HMZ	North Korea
CYA-CZZ	Canada	HNA-HNZ	Iraq
C2A-C2Z	Nauru	HOA-HPZ	Panama
C3A-C3Z	Andorra	HOA-HRZ	Honduras
C4A-C4Z	Cyprus	HSA-HSZ	Thailand
C5A-C5Z	The Gambia	HTA-HTZ	Nicaragua
C6A-C6Z	Bahamas	HUA-HUZ	El Salvador
C7A-C7Z	World Met. Org.	HVA-HVZ	Vatican City State
C8A-C9Z	Mozambique	HWA-HYZ	France
DAA-DRZ	Germany	HZA-HZZ	Saudi Arabia
DSA-DTZ	South Korea	H2A-H2Z	Cyprus
DUA-DZZ	Philippines	H3A-H3Z	Panama
D2A-D3Z	Angola	H4A-H4Z	Solomon Islands
D4A-D4Z	Cape Verde	H6A-H7Z	Nicaragua
D5A-D5Z	Liberia	H8A-H9Z	Panama
D6A-D6Z	Comoros	IAA-IZZ	Italy
D7A-D9Z	South Korea	JAA-JSZ	Japan
EAA-EHZ	Spain	JTA-JVZ	Mongolian People's Rep.

Call Sign	Allocated to	Call Sign	Allocated to
JWA-JXZ	Norway	SSA-SSM	Egypt
JYA-JYZ	Jordan	SSN-STZ	Sudan
JZA-JZZ	Indonesia	SUA-SUZ	Egypt
J2A-J2Z	Djibouti	SVA-SZZ	Greece
J3A-J3Z	Grenada	S2A-S3Z	Bangladesh
J4A-J4Z	Greece	S6A-S6Z	Singapore
J5A-J5Z	Guinea-Bissau	S7A-S7Z	Seychelles
J6A-J6Z	Saint Lucia	S9A-S9Z	Sao Tome and Principe
J7A-J7Z	Dominica	TAA-TCZ	Turkey
J8A-J8Z	St Vincent and the Grenadines	TDA-TDZ	Guatemala
		TEA-TEZ	Costa Rica
KAA-KZZ	United States of America	TFA-TFZ	Iceland
LAA-LNZ	Norway	TGA-TGZ	Guatemala
LOA-LWZ	Argentine Republic	THA-THZ	France
LXA-LXZ	Luxembourg	TIA-TIZ	Costa Rica
LYA-LYZ	Russian Federation	TJA-TJZ	Cameroon
LZA-LZZ	Bulgaria	TKA-TKZ	France
L2A-L9Z	Argentine Republic	TLA-TLZ	Central African Republic
MAA-MZZ	United Kingdom	TMA-TMZ	France
NAA-NZZ	United States of America	TNA-TNZ	Congo
OAA-OCZ	Peru	TOA-TQZ	France
ODA-ODZ	Lebanon	TRA-TRZ	Gabonese Republic
OEA-OEZ	Austria	TSA-TSZ	Tunisia
OFA-OJZ	Finland	TTA-TTZ	Chad
OKA-OMZ	Czech & Slovak Federal Rep.	TUA-TUZ	Ivory Coast
		TVA-TXZ	France
ONA-OTZ	Belgium	TYA-TYZ	Benin
OUA-OZZ	Denmark	TZA-TZZ	Mali
PAA-PIZ	Netherlands	T2A-T2Z	Tuvalu
PJA-PJZ	Nederlands Antillen	T3A-T3Z	Kiribati
PKA-POZ	Indonesia	T4A-T4Z	Cuba
PPA-PYZ	Brazil	T5A-T5Z	Somali Democratic Rep.
PZA-PZZ	Suriname	T6A-T6Z	Afghanistan
P2A-P2Z	Papua New Guinea	T7A-T7Z	San Marino
P3A-P3Z	Cyprus	UAA-UAZ	Russian Federation
P4A-P4Z	Aruba	URA-UTZ	Ukraine
P5A-P9Z	North Korea	UUA-UZZ	Russian Federation
RAA-RZZ	Russian Federation	VHA-VNZ	Australia
SAA-SMZ	Sweden	VOA-VOZ	Canada (Also VAA-VGZ)
SNA-SRZ	Poland	VPA-VSZ	United Kingdom

Call Sign	Allocated to	Call Sign	Allocated to
VTA-VWZ	India	ZPA-ZPZ	Paraguay
VXA-VYZ	Canada	ZOA-ZQZ	United Kingdom
VZA-VZZ	Australia	ZRA-ZUZ	South Africa
V2A-V2Z	Antigua and Barbuda	ZVA-ZZZ	Brazil
V3A-V3Z	Belize	Z2A-Z2Z	Zimbabwe
V4A-V4Z	Saint Kitts and Nevis	2AA-2ZZ	United Kingdom
V5A-V5Z	Namibia	3AA-3AZ	Monaco
V6A-V6Z	Micronesia	3BA-3BZ	Mauritius
V7A-V7Z	Marshall Islands	3CA-3CZ	Equatorial Guinea
V8A-V8Z	Brunei Darussalem	3DA-3DM	Swaziland
WAA-WZZ	United States of America	3DN-3DZ	Fiji
XAA-XIZ	Mexico	3EA-3FZ	Panama
XJA-XOZ	Canada	3GA-3GZ	Chile
XPA-XPZ	Denmark	3HA-3UZ	China
XQA-XRZ	Chile	3VA-3VZ	Tunisia
XSA-XSZ	China	3WA-3WZ	Viet Nam
XTA-XTZ	Burkina Faso	3XA-3XZ	Guinea
XUA-XUZ	Cambodia	3YA-3YZ	Norway
XVA-XVZ	Viet Nam	3ZA-3ZZ	Poland
XWA-XWZ	Lao People's Dem. Rep.	4AA-4CZ	Mexico
XXA-XXZ	Portugal	4DA-4IZ	Philippines
XYA-XZZ	Myanmar	4JA-4LZ	Russian Federation
YAA-YAZ	Afghanistan	4MA-4MZ	Venezuela
YBA-YHZ	Indonesia	4NA-4OZ	Yugoslavia (former)
YIA-YIZ	Iraq	4PA-4SZ	Sri Lanka
YJA-YJZ	Vanuatu	4TA-4TZ	Peru
YKA-YKZ	Syria	4UA-4UZ	United Nations Org.
YLA-YLZ	Russian Federation	4VA-4VZ	Haiti
YMA-YMZ	Turkey	4XA-4XZ	Israel
YNA-YNZ	Nicaragua	4YA-4YZ	Int. Civil Aviation Org.
YOA-YRZ	Romania	4ZA-4ZZ	Israel
YSA-YSZ	El Salvador	5AA-5AZ	Libya
YTA-YUZ	Yugoslavia (former)	5BA-5BZ	Cyprus
YVA-YYZ	Venezuela	5CA-5GZ	Morocco
YZA-YZZ	Yugoslavia (former)	5HA-5IZ	Tanzania
Y2A-Y9Z	Germany	5JA-5KZ	Colombia
ZAA-ZAZ	Albania	5LA-5MZ	Liberia
ZBA-ZJZ	United Kingdom	5NA-5OZ	Nigeria
ZKA-ZMZ	New Zealand	5PA-5QZ	Denmark
ZNA-ZOZ	United Kingdom	5RA-5SZ	Madagascar

Call Sign	Allocated to	Call Sign	Allocated to
5TA-5TZ	Mauritania	7TA-TYZ	Algeria
5UA-5UZ	Niger	7ZA-7ZZ	Saudi Arabia
5VA-5VZ	Togolese Republic	8AA-8IZ	Indonesia
5WA-5WZ	Western Samoa	8JA-8NZ	Japan
5XA-5XZ	Uganda	8QA-8QZ	Botswana
5YA-5ZZ	Kenya	8PA-8PZ	Barbados
6AA-6BZ	Egypt	8OA-8OZ	Maldives
6CA-6CZ	Syria	8RA-8RZ	Guyana
6DA-6JZ	Mexico	8SA-8SZ	Sweden
6KA-6NZ	Korea	8TA-8YZ	India
6OA-6OZ	Somali Dem. Rep.	8ZA-8ZZ	Saudi Arabia
6PA-6SZ	Pakistan	9BA-9DZ	Iran
6TA-6UZ	Sudan	9EA-9FZ	Ethiopia
6VA-6WZ	Senegal	9GA-9GZ	Ghana
6XA-6XZ	Madagascar	9HA-9HZ	Malta
6YA-TYZ	Jamaica	9IA-9JZ	Zambia
6ZA-6ZZ	Liberia	9KA-9KZ	Kuwait
7AA-7IZ	Indonesia	9LA-9LZ	Sierra Leone
7JA-7NZ	Japan	9MA-9MZ	Malaysia
7QA-7QZ	Yemen	9NA-9NZ	Nepal
7PA-7PZ	Lesotho	9OA-9TZ	Zaire
7OA-7OZ	Malawi	9UA-9UZ	Burundi
7RA-7RZ	Algeria	9VA-9XZ	Singapore
7SA-7SZ	Sweden	9WA-9WZ	Malaysia
		9XA-9XZ	Rwanda
		9YA-9ZZ	Trinidad and Tobago

Appendix C

MARITIME MOBILE FREQUENCY ALLOCATIONS

The Maritime Mobile Service is allocated specific frequencies, usually together in various 'bands' throughout the radio-frequency spectrum. The main bands and frequencies of interest to mariners are shown below – together with typical use.

Frequency/Band	Services/Facilities Provided
415–535kHz	Radiotelegraph (Morse) Band, including 500kHz Morse Distress and Calling frequency (to become redundant in most areas with the advent of the GMDSS). 518kHz NAVTEX (to become part of the GMDSS). Morse Weather and Navigation Warnings, and Telegram Services.
1605–4000kHz	Main 2MHz Telephony and NBDP (SITOR/Telex) band. Includes 2182kHz Voice Distress and Calling frequency, and 2187.5kHz DSC frequency. Radiotelephone and Radiotelex service via Coast Stations. Weather and Navigation Warning Broadcasts. Inter-ship Voice channels.
4MHz–27.5MHz	High Frequency/Long Range Radiotelephone bands, for the Public Correspondence Telephone Service. HF DSC frequencies for Distress Alerting. Weather and Navigation Warnings. Inter-ship frequencies. The maritime bands are interleaved with many other services across the HF spectrum.
121.5MHz	Civil Aircraft EPIRB frequency. Can be used to 'home-in' on a casualty, by SAR aircraft.
156–163MHz	VHF Radiotelephone Service. Port, Harbour and Vessel Traffic Services. Inter-ship. Local Weather and Navigation Warnings. Includes CH16 – the primary short range voice distress and calling channel, for use also for on-scene communications in emergency situations. CH70 – VHF DSC channel for distress alerting within the GMDSS.
243.0MHz	Military aircraft distress frequency – used by EPIRB's (usually in conjunction with 121.5MHz).
216–220MHz	Automated Maritime Telephone Services (USA) – Public Correspondence Telephone Service .
406–406.1MHz	406MHz satellite EPIRB. A cornerstone of the GMDSS, and the only EPIRB which allows world-wide distress alerting and position fixing.
457.525–467.825MHz	Onboard UHF
1530–1559MHz 1625.5–1660.5MHz	Maritime Mobile Satellite Service. Some frequencies are 'generic' and can be used by non-Maritime Mobiles. New generic frequencies now also available to maritime users, for which equipment will have to be developed.
9300-9500MHz	Radar - within which band the 'Search and Rescue Transponder' (SART) will operate, to attract help to the position of a casualty

MARINE VHF INTERNATIONAL CHANNEL ALLOCATIONS & NATIONAL DIFFERENCES

The short-range nature of VHF makes it ideal for local communications (eg, within the limits of a port). There is also a facility for nations to use some frequencies in a different way to the international allocation, as summarised below. Whilst these differences do not prevent the use of a VHF set internationally, mariners with equipment which does not have all channel options may find themselves unable to receive vital information, nor to communicate efficiently with some coast stations.

CHANNEL DESIG- NATOR	TRANSMITTING FREQ(MHZ) SHIP	COAST	INTERNATIONAL USE INTER- SHIP	PORTS &VTS	PUBLIC CORRE SP'NCE	NATIONAL PECULIARITIES (See notes) USA	CANADA	UK
60	156.025	160.625		X	X			
01	156.050	160.650		X	X			
61	156.075	160.675		X	X			
02	156.100	160.700		X	X			
62	156.125	160.725		X	X			
03	156.150	160.750		X	X			
63	156.175	160.775		X	X			
04	156.200	160.800		X	X			
64	156.225	160.825		X	X			
05	156.250	160.850		X	X	05A	05A	
65	156.275	160.875		X	X	65A		
06	156.300	-	X					
66	156.325	160.925		X	X	66A		
07	156.350	160.950		X	X	07A		
(Bottom range channels are mainly Duplex; mid-range are Simplex)								
67	156.375	156.375	X	X				CG
08	156.400	-	X					
68	156.425	156.425		X				
09	156.450	156.450	X	X				
69	156.475	156.475	X	X				
10	156.500	156.500	X	X				
70	156.525	156.525 - *DIGITAL SELECTIVE CALLING (GMDSS)*						
11	156.550	156.550		X				
71	156.575	156.575		X				
12	156.600	156.600		X				
72	156.625	-	X					
13	156.650	156.650	X	X				
73	156.675	156.675	X	X				
14	156.700	156.700		X				

CHANNEL DESIG-NATOR	TRANSMITTING FREQ(MHZ)		INTERNATIONAL USE			NATIONAL PECULIARITIES (See notes)		
	SHIP	COAST	INTER-SHIP	PORTS &VTS	PUBLIC CORRE SP'NCE	USA	CANADA	UK
74	156.725	156.725		X				
15	156.750	156.750	X	X				
75	(GUARD BAND FOR CHANNEL 16 DISTRESS & SAFETY CHANNEL)							
16	**156.800**	**156.800 - DISTRESS, SAFETY AND CALLING CHANNEL**						
76	(GUARD BAND FOR CHANNEL 16 DISTRESS & SAFETY CHANNEL)							
17	156.850	156.850	X	X				
77	156.875	-	X					

(Mid-range channels are Simplex; Top-range are Duplex)

CHANNEL DESIG-NATOR	SHIP	COAST	INTER-SHIP	PORTS &VTS	PUBLIC CORRE SP'NCE	USA	CANADA	UK
18	156.900	161.500		X		18A		
78	156.925	161.525		X	X	78A		
19	156.950	161.550		X		19A		
79	156.975	161.575		X		79A		
20	157.000	161.600		X				
80	157.025	161.625		X		80A		M1
21	157.050	161.650		X		21CG	21B	
81	157.075	161.675		X	X			
22	157.100	161.700		X		22CG	22A	
82	157.125	161.725		X	X			
23	157.150	161.750			X	23CG		
83	157.175	161.775			X	83CG	83B	
24	157.200	161.800			X			
84	157.225	161.825	X		X			
25	157.250	161.850			X		25B	
85	157.275	161.875			X			
26	157.300	161.900			X			
86	157.325	161.925			X			
27	157.350	161.950			X			
87	157.375	161.975			X			
28	157.400	162.000			X			
88	157.425	162.025			X	88A		
37	157.850	157.850						M
	161.425	161.425						M2
		162.400				WX2	WX2	
		162.475				WX3	WX3	
		162.550				WX1	WX1	

Notes:

1 In Canada and the USA, channels designated 'A' are Simplex versions of the Duplex numbered-channel — with the ship transmit frequency being used also by the Coast Station concerned. VHF sets with only 'International' channels cannot communicate on these channels, unless a 'private' channel has been programmed with this (simplex) frequency.

2 In Canada and the USA, channels designated 'B' are Broadcast channels, using the normal Coast Station frequency for transmission. VHF sets with only international channels can receive these broadcasts. Coast Stations will not normally monitor the Ship Transmit frequency.

3 Channels WX1 to WX3 are broadcast channels, sending continuous weather (USA) and weather and navigational information (Canada), where used. VHF sets need the 'USA' option to receive these channels, or have a 'private' channel programmed for this purpose. The USA National Oceanic & Atmospheric Administration operates a network of stations which can be received when in USA waters. Continuous Weather/Navigational broadcasts in Canadian waters mainly use channels 21B, 83B and 25B — with WX1 being used by some Pacific Coast stations.

The US Coast Guard uses Channel 22CG for scheduled weather and navigation broadcasts (22CG, like 21CG and 23CG, use the ship transmit frequency as a simplex channel — 'USA' channelisation/private channel programming required to work these channels and to receive UC Coast Guard broadcasts on 22CG).

Other stations (eg. Coast Radio Stations and Port authorities) also broadcast weather and navigational information, as happens in many countries internationally, at scheduled times on their normal working channels.

4 Channel 6 — the primary Inter-ship channel (Internationally) is designated as 'Inter-ship Safety' in the USA. The channel should only be used for inter-ship safety communications, and for SAR communications with US Coast Guard vessels and aircraft, when in US waters.

5 Channel 80 is the primary marina channel in the UK. VHF sets with USA channels only cannot communicate with UK marinas on this (Duplex) channel (International channelisation is required).

6 Channel 67 is the 'Small Ship Safety' channel in the UK — for communication with the Coastguard, after initial contact on Channel 16.

HF RADIOTELEPHONE STATIONS

There is a large number of HF radio stations world-wide, providing a selection of Public Correspondence and safety services (DSC watch, weather and navigation warning broadcasts etc) — by voice, radio telex, Morse and facsimile.

The following sample of radiotelephone stations (two in each of the three ITU Regions) is provided to allow readers to listen-in to HF R/T stations: to become familiar with the calling procedures, traffic list and safety broadcast routines. With a reasonable receiver/antenna arrangement, at least one of the six stations listed should be in range of virtually any position on the globe, at any time of day or night. (All frequencies in kilohertz - kHz).

REGION 1 - Europe, Africa and the Middle East

PORTISHEAD RADIO/GKA

ITU CHAN	PORTISHEAD CALLSIGN	SHIP FREQ. RECEIVE	TRANSMIT	REMARKS	ITU CHAN	PORTISHEAD CALLSIGN	SHIP FREQ. RECEIVE	TRANSMIT	REMARKS
410	GKT20	4384	4092	(1)	1602	GKT62	17245	16363	(1)
402	GKT22	4360	4068		1606	GKT66	17257	16375	
406	GKT26	4372	4080		1611	GKU61	17272	16390	
426	GKV26	4432	4140		1615	GKU65	17284	16402	
					1618	GKU68	17293	16411	
816	GKU46	8764	8240	(1)	1623	GKV63	17308	16426	
802	GKT42	8722	8198		1632	GKW62	17335	16453	
819	GKU49	8773	8249		1637	GKW67	17350	16468	
822	GKU42	8782	8258		1640	GKW60	17359	16477	
826	GKV46	8794	8270						
831	GKW41	8809	8285		1801	GKT18	19755	18780	(1)
					1803	GKU18	19761	18786	
1224	GKV54	13146	12299	(1)					
1201	GKT51	13077	12230		2206	GKT76	22711	22015	(1)
1202	GKT52	13080	12233		2212	GKU72	22729	22033	
1206	GKT56	13092	12245		2214	GKU74	22735	22039	
1228	GKV58	13158	12311		2220	GKU70	22753	22057	
1230	GKV50	13164	12317		2227	GKV77	22774	22078	
1232	GKW52	13170	12323		2229	GKV79	22780	22084	
					2240	GKX70	22813	22117	
					2502	GKU25	26148	25073	(1)

(1) Main Channel: Traffic lists are broadcast on these channels — every hour, on the hour. Portishead maintains a listening watch on Main Channels, for calls from vessels requiring manual connection.

Initial calls procedure:

'Portishead Radio (three times) this is (vessel name - three times, and callsign) calling you on GKV54 (or other main channel).'

Portishead will give you a working channel. You should shift to the new channel and repeat the call, unless you were told to 'standby', in which case Portishead will call you on that channel when ready.

If you do not get a reply to your initial call, wait a few minutes and try again. If there is no response after a few calls, shift bands and try again.

(For full details of Portishead Radio services, write to the address in Appendix Q)

Cape Town Radio/ZSC										
ITU CHAN	ZSC CHAN DESIG'	SHIP FREQ.		REMARKS		ITU CHAN	ZSC CHAN DESIG'	SHIP FREQ.		REMARKS
		RECEIVE	TRANSMIT					RECEIVE	TRANSMIT	
-	ZSC56	4125	4125			1209	ZSC27	13101	12254	
405	ZSC25	4369	4077			1221	ZSC54	13137	12290	H24
421	ZSC53	4417	4125	H24						
424	ZSC48	4426	4134			1608	ZSC28	17263	16381	
427	ZSC67	4435	4143	(2)		1621	ZSC59	17302	16420	H24
						1633	-	17338	16456	
805	ZSC26	8731	8207							
821	ZSC57	8779	8255	H24		2204	ZSC29	22705	22009	
						2206	-	22711	22015	
						2221	ZSC60	22756	22060	H24

Note 2: Traffic lists on 4435kHz (ITU Chan 427) at 0003, 0603, 0948, 1403, 1748 UTC.

Storm Warnings on 4435kHz at the end of the first Silence Period after receipt (0003, 0033 etc).

Gale Warnings and reports from Met Obs stations on 4435kHz at 1333 UTC.

Gale Warnings and main Weather Forecast on 4435kHz for Cape coastal waters, and Cape West area.

Urgent Navarea VII, and urgent coastal warnings for Cape coastal waters, on 4435kHz at the end of the first Silence Period after receipt at ZSC.
Public Correspondence (Link Calls) — call direct on one of the H24 channels (after ensuring it is free) and be prepared to move to another working channel.

REGION 2 - THE AMERICAS

In Region 2, Canada and the USA are particularly well served with HF stations. The Canadian Coastguard operates two stations — one on the Atlantic and one on the Pacific coast. The USCG provides a service, and a number of US companies also operate stations across the continent.

USA - PACIFIC COAST AT & T Station KMI (California)			

ITU CHAN	SHIP FREQUENCY			ITU CHAN	SHIP FREQUENCY		
	RECEIVE	TRANSMIT	REMARKS		RECEIVE	TRANSMIT	REMARKS
242	2450	2003		1201	13077	12230	
248	2506	2406		1202	13080	12233	
				1203	13083	12236	(1)
401	4357	4065					
416	4402	4110	(1)	1229	13161	12314	
417	4405	4113		1602	17245	16363	
				1603	17248	16366	
804	8728	8204		1624	17311	16429	
809	8743	8219					
822	8782	8258		2214	22735	22039	
				2223	22762	22066	
				2228	22777	22081	
				2236	22801	22105	

(1) Broadcast Channels: Traffic lists sent every four hours, starting midnight. UTC. Weather forecast at 0000 and 1600 UTC. AT&T stations monitor all channels, not just the 'main' channel in any one band.

Callers must state which channel they are calling on, eg:
'KMI Channel Four One Seven, KMI Channel Four One Seven, KMI Channel Four One Seven. This is yacht *Happy Daze* off San Diego, Over'

(For full details of AT&T maritime radio services write to the address in Appendix Q, or contact any of the three AT&T stations in New Jersey, Florida or California).

CANADA - ATLANTIC COAST
Halifax Radio/VCS

ITU CHAN	SHIP FREQUENCY		REMARKS	ITU CHAN	SHIP FREQUENCY		REMARKS
	RECEIVE	TRANSMIT			RECEIVE	TRANSMIT	
418	4408	4116		1213	13113	12266	
605	6513	6212		1604	17251	16369	
823	8785	8261		2213	22732	22036	

Operating on one channel only in each band, the Canadian CG provides a safety and public correspondence service.

Traffic lists on all channels at 0133, 0533, 0933, 1333, 1733, 2133 UTC. Ice Reports (Gulf of St Lawrence) and Iceberg Bulletins on 4408, 8785 and 13113 at 1335 and 2335 UTC. Weather forecast on all channels at 0205, 0805, 1605, 2205. Navigation warnings on all channels at 0335 0735 1535 2135.

REGION 3 - AUSTRALISIA, SE ASIA, W. PACIFIC

South East Asia is served by a large number of HF R/T stations — many operating on only a few bands. Australia (Telstra Maritime) operates five HF stations, two listed here.

Perth Radio/VIP/Selcall 0107							
ITU CHAN	SHIP FREQUENCY		REMARKS	ITU CHAN	SHIP FREQUENCY		REMARKS
	RECEIVE	TRANSMIT			RECEIVE	TRANSMIT	
404	4366	4074		-	12365	-	(1)
415	4399	4107					
424	4426	-	(1)	1226	13152	12305	V(24H)TS
427	4143	4435	V(H24)TS	1229	13161	12314	TS
603	6507	-	(1)	1604	17251	16369	V(2200-1400)TS
607	6519	6218	TS	1610	17269	16387	TS
				1612	17275	16393	
-	8176	-	(1)				
806	8734	8210	V(24H)TS	2212	22729	22033	V(2200-1400)T
811	8749	8225		2228	22777	22081	
815	8761	8237		2502	26148	25073	

(1) Coastal Weather reports for South and West Coasts (W Australia) at 2318(0718), 1318(0518), and 1918(1118) UTC (Local Standard Time). Western Area High Seas forecast at 0118(0918) and 1518(2318).

ITU CHAN	SHIP FREQUENCY		REMARKS	ITU CHAN	SHIP FREQUENCY		REMARKS
	RECEIVE	TRANSMIT			RECEIVE	TRANSMIT	
405	4369	4077	V(H24)TS	-	12365	-	(2)
417	4405	4113					
424	4426	-	(2)	1203	13083	12236	V(H24)TS
				1231	13167	12320	
603	6507	-	(2)				
607	6519	6218	V(H24)TS	1602	17245	16363	V(H24)TS
				1610	17269	16387	TS
-	8176	-	(2)	1622	17305	16423	
802	8722	8198	V(H24)TS				
829	8803	8279		2203	22702	22006	V(2000-0900)T
				2223	22762	22066	
				2502	26148	25073	

Sydney Radio/VIS/Selcall 0108

(2) Coastal Weather reports for Queensland/NSW at 2133(0533), 0533(1333) and 0933(1733) UTC(LST). South-East area High Seas weather at 0333(1103) and 1333(2103).

Navarea 10 navigation warnings are broadcast with the High Seas weather reports. 'Auscoast' warnings are broadcast with the Coast Weather Reports.

In case of maritime distress/emergency, call on CH16 VHF, 2182kHz or on 4125, 6215, or 8291kHz; and/or 12290/16420kHz (0700-1900 local).

Public Correspondence: Both stations are open 24 hours. Call on the first frequencies indicated as, eg V(H24)TC using voice, Tone-call or Selcall as appropriate. You will then be allocated a working channel. Traffic lists are broadcast every hour, on the hour, on those same voice calling channels (if free), and Channel 834.

For full details of Telstra Maritime services, contact the address in Appendix Q.

Appendix F

SHIP-TO-SHIP VOICE CHANNELS (MF & HF SSB)

A number of frequencies have been provided for 'World-wide common use by ships of all categories' which means that any suitably licenced vessel/craft in the Maritime Mobile Service can make use of them. The frequencies are not assigned to any particular ship/coast station. When working ship-to-ship, each frequency can be used for simplex (one-frequency, two-way communication) or cross-band (duplex) working. The 4MHz and 8MHz frequencies may also be used for duplex working with Coast Stations, (normally as directed by the Coast Station).

HF FREQUENCIES

Carrier Frequency	Assigned Frequency	Carrier Frequency	Assigned Frequency	Carrier Frequency	Assigned Frequency
4146	4147.4	16528	16529.4	22159	22160.4
4149	4150.4	16531	16532.4	22162	22163.4
		16534	16535.4	22165	22166.4
6224	6225.4	16537	16538.4	22168	22169.4
6227	6228.4	16540	16541.4	22171	22172.4
6230	6231.4	16543	16544.4	22174*	22175.4
		16546	16547.4	22177*	22178.4
8294	8295.4				
8297	8298.4	18825*	18826.4	25100*	25101.4
		18828*	18829.4	25103*	25104.4
		18831*	18832.4	25106*	25107.4
		18834*	18835.4	25109*	25110.4
* Restricted to Public		18837*	18838.4	23112*	25113.4
Correspondence use		18840	18841.4	25115	25116.5
in USA		18843	18844.4	25118	25119.5

You will remember from Chapter 2 that HF propagation is primarily used for "long range" communication, using sky wave propagation. The above frequencies can be used in that way to cover thousands of miles. However, there is also the ground wave cover to consider — and this can stretch from a few miles only at the higher bands, to over 100 miles at 4MHz, during the day. When working vessels close to yourself, you should reduce power to the minimum required to cover the distance.

MF INTER-SHIP FREQUENCIES

Your licensing authority will tell you which (MF/HF) inter-ship frequencies you are allowed to use (You may have to ask). As a guide — the bands 2262.5 — 2498kHz and 3340 — 3400kHz have been set aside for inter-ship, SSB radiotelephony in Region 1.

In Regions 2 and 3, 2635kHz and 2638kHz are provided in addition to locally prescribed frequencies.

All frequencies shown are for SSB (USB) operation. The Carrier Frequency is the one which you will dial-up on a Marine SSB TXR. The Assigned Frequency is the centre-frequency of the transmitted sideband. If you want to monitor these frequencies using radio equipment with different offsets to Marine SSB, you should work back from the assigned frequency to find the proper settings.

FREQUENCY STANDARD AND TIME SIGNAL STATIONS

Older radio receivers have to be 'calibrated' regularly – each time you change band is good practice. To help you calibrate, a number of stations broadcast on frequencies across the bands – and with nice round numbers (see table below). The frequencies double-up as time signals and may also include other information (eg, weather warnings). Modern, synthesized receivers rarely drift in frequency when in use – but it does no harm to convince yourself that your set is accurate.

	Stations sharing key standard frequencies (kHz)						
Station & Location	2500	5000	8000	10000	15000	16000	20000
Fort Collins **(1)** Colorado - WWV	H24	H24	-	H24	H24	-	H24
Kekaha **(1)** Hawaii - WWVH	H24	H24	-	H24	H24	-	-
Station JJY **(2)** Tokyo Japan	H24	H24	H24	H24	H24	-	-
LOL **(3)** Buenos Aires	-	(3)	-	(3)	(3)	-	-
VNG Llandilo, Australia	H24	H24	-	-	-	2200 - 1000	-
BSF Taiwan	-	H24	-	-	H24	-	-

(1) Fort Collins (WWV) and KEKAHA (WWVH) transmit voice time announcements every minute – a male voice for WWV and a female voice for WWVH. Time announcements in the form 'At the tone, 14 minutes, Co-ordinated Universal Time'. The WWV voice starts around 7.5 seconds before the tone and is preceded by the WWVH announcement. WWV provides brief weather announcements starting at H+08 – including information on the movement of storm centres, as well as Omega bulletins, covering the NW Atlantic, Gulf of Mexico and the Caribbean Sea. WWVH provides similar information for the Pacific Ocean, starting at H+48. Bulletins from both stations last around 3 minutes only.
(2) Station JJY makes announcements of Japan Standard Time in slow Morse code and by voice.
(3) LOL, Buenos Aires hours of service (UTC) are: 1100-1200; 1400-1500; 1700-1800; 2000-2100; 2300-2400.

Another station worth monitoring is CHU, Ottawa, Canada. Operating on 3330,7335 and 14670kHz, you avoid the problems of separating one station from another. Announcements are made in French and English and transmissions are continuous.

WWV (Fort Collins) and Rugby Radio (England) both broadcast continuously on 60kHz – which you may be able to get on your receiver even if it states its frequency range as being 150kHz-28MHz. Receiver coverage claimed will only include the range which falls within stated levels of selectivity and sensitivity, so receiver range, albeit at reduced performance, often stretches beyond that stated in manufacturers literature. The time signal is in the form of a brief coded message which automatically sets/resets special clocks with a built-in 60kHz receiver.

WEATHERFAX STATIONS

For equipment requirements to receive Weatherfax Broadcasts, see Chapter 6. A large number of stations world-wide send weatherfax broadcasts. The following stations have been selected as examples. Most transmit a 'schedule' to let you know what to expect, on what frequency and at what time. Marine SSB Receiver settings(USB mode) shown.

Channel Desig.	Centre Freq. (KHz)	SSB RX Setting	Operating Times (UTC)	Notes
Australia - Darwin (AXI) - Broadcast Schedule - 0030				
AX 132	5755	5753.1	1100 - 2300	Covers much of SE Asia, Australia and
AX 133	7535	7533.1	1100 - 2300	New Zealand waters.
AX 134	10555	10553.1	H24	Frequency Recommendations are broadcast
AX 135	15615	15613.1	2300 - 1100	at 0000.
AX 137	18060	18058.1	2300 - 1100	
Australia - Canberra (AXM) - Broadcast Schedule - 0115				
AXM 31	2628	2626.1	H24	Covers S Hemisphere from 0°, east to 180°.
AXM 33	5100	5098.1	H24	
AXM 34	11030	11028.1	H24	Frequency Recommendations - 0130
AXM 35	13920	13918.1	H24	
AXM 37	20469	20467.1	H24	
Canada - Esquimalt (CKN) - Broadcast Schedule - 0245				
	2752.1	2750.2	H24	Covers W Canada Coast and NE Pacific waters
	4266.1	4264.2	H24	to around 170°W.
	6454.4	6452.5	H24	
	12751.1	12749.2	H24	
Canada - Halifax (CFH) - Broadcast Schedule - 1014 (Mon,Wed,Fri,Sun)				
	122.5	120.6	H24	Covers Canada & USA East Coast, Caribbean, &
	4271	4269.1	2200 - 1000	NW Atlantic.
	6496.4	6494.5	H24	Test chart - 1014 (Tue, Thu, Sat)
	10536	10534.1	H24	Frequencies shared with Teletype Broadcasts.
	13510	13508.1	1000 - 2200	
Canada - Iqaluit (VFF) and Resolute (VFR) (Frequencies on time-share basis)				
	3253	3251.1	H24	Covers Atlantic Coast; Gulf & River St
	7710	7708.1	H24	Lawrence to Montreal; Eastern Arctic; Hudson Bay & Strait. (Areas covered in broadcasts vary with ice conditions and level of maritime activity).
USA - Mobile (WLO) Alabama - Radiofax Schedule - 1500				
	2572	2570.1	1200 - 2400	Gulf Surface Analysis @ 0230,0920,1430,2020
	6852	6850.1	H24	USA Surface Analysis @ 0735,1910
	9157.5	9155.6	1200 - 2400	High Seas Forecast @ 1100,2300 18/36 hour Gulf Surface Prognosis 1145,2345

Channel Desig.	Centre Freq. (KHz)	SSB RX Setting	Operating Times (UTC)	Notes
England	**Bracknel**	**(GFA and GFE)**		
GFA 21	3289.5	3287.6	H24	Radio frequency check -1400
GFA 22	4610	4608.1	1800 - 0600	General Notices -1622
GFA 23	8040	8038.1	H24	Covers most of North Atlantic, Baltic and
GFA 24	11086.5	11084.6	H24	Med for Surface analysis and prognosis, Sea-
GFA 25	14582.5	14580.6	0600 - 1800	State, Inference etc.
GFE 25	2618.5	2616.6	1800 - 0600	October - March
			1900 - 0500	April - September
GFE 21	4782	4780.1	H24	
GFE 22	9203	9201.1	H24	
GFE 24	18261	18259.1	0600 - 1800	October - March
			0500 - 1900	April - September

England - Northwood - Broadcast Schedule - 0300, 1640

	2374	2372.1	1630 - 0730 (Oct-Mar)	Covers much of N Atlantic from about 30°N (excluding East Coast USA) up to Arctic
	3652	3650.1	H24	region (approx 70°N). Surface analysis,
	4307	4305.1	H24	Sea State prognosis, Gale Warnings etc.
	6446	6444.1	H24	
	8331.5	8329.6	H24	
	12844.5	12842.6	H24	
	16912	16910.1	H24	April - September
			0730 -1630	October - March

Italy - Roma (IMB) - Test Chart - 0813

IMB 51	4777.5	4775.6	0400 - 2400	Mediterranean Sea state prognosis, surface
IMB 55	8146.6	8144.7	0400 - 2400	analysis, etc.
IMB 56	13597.4	13595.5	0400 - 2400	

South Africa - Pretoria (ZRO) - Broadcast Schedule - 0505

ZRO 5	4014	4012.1	H24	Covers SE Atlantic, SW Indian Ocean, and
ZRO 2	7508	7506.1	H24	coastal waters of South Africa. Surface
ZRO 3	13538	13536.1	H24	prognosis/analysis, etc - and northern
ZRO 4	18238	18236.1	H24	ice limits (Antartica)

Honolulu (KVM 70)

	9982.5	9980.6		Covers most of the Pacific Ocean and East
	11090	11088.1		Indian Ocean, in varying scales.
	16135	16133.1		
	23331.5	23329.6		

Buenos Aires (LRB/LRO) Broadcast Schedule - 1430

LRO 69	5185	5183.1	H24	Covers area bounded by 7S.95W; 5S.35W;
LRB 72	10720	10718.1	H24	40S.144W; 39S.14W.
LRO 84	18093	18091.1	H24	

NAVTEX STATIONS AND BROADCAST TIMES WORLDWIDE

As the GMDSS is progressively introduced, the number of NAVTEX stations world-wide is expected to increase. The table below shows the known situation, including some future planned stations, at the time of going to print with this edition of the manual. All broadcasts are in the English language, except where shown differently. Broadcasts are on the international NAVTEX frequency of 518kHz, unless otherwise stated.

LOCATION	STATION	STATION IDENT	BROADCAST TIMES						REMARKS
NAVAREA I									
N Russia	ARKHANGELSK	F	0200	0600	1000	1400	1800	2200	
N Russia	MURMANSK	C	0120	0520	0920	1320	1720	2120	
Iceland	REYKJAVIK	R	0318	0718	1118	1518	1918	2318	Pre-Operational
N Norway	VARDO	V	0300	0700	1100	1500	1900	2300	
W Norway	BODO	B	0018	0418	0900	1218	1618	2100	
SW Norway	ROGALAND	L	0148	0548	0948	1348	1748	2148	
W Sweden	STOCKHOLM	J	0300 0730 1930 1130 1530 2330						Weather Forecast Ice Report
NW Sweden	HAERNOSAND	H	0000 0400 0800 1600 2000 1200						Weather Forecast Ice Report
Estonia	TALLINN	U	0030 0430 0830 1630 2030 1230						Weather Forecast Ice Report
Netherlands	CG IJMUIDEN	P	0348	0748	1148	1548	1948	2348	CG=Coastguard
Belgium	OOSTENDE	T	0248	0648	1048	1448	1848	2248	
W England	CULLERCOATS	G	0048	0448	0848	1248	1648	2048	
Isle of Wight	NITON	S	0018	0418	0818	1218	1618	2018	
SW Scotland	PORTPATRICK	O	0130	0530	0930	1330	1730	2130	
SW Ireland	VALENTIA	(TBD)	(TO BE DECIDED)						
NW France	BREST LE CONQUET	F							
NAVAREA II									
NW France	CROSS CORSEN	A	0000	0400	0800	1200	1600	2000	
W Spain	FINISTERRE	D	0030	0430	0830	1230	1630	2030	(Trial-also in Spanish)
Portugal	LISBON	R	0250	0650	1050	1450	1850	2250	
Azores	HORTA	F	0050	0450	0850	1250	1650	2050	
Canary Isles	CANARY ISLES	I	0100	0500	0900	1300	1700	2100	(As FINISTERRE)
Cameroon	DUALA	(TBD)	(TO BE DECIDED)						

LOCATION	STATION	STATION IDENT	BROADCAST TIMES						REMARKS
NAVAREA III									
S Spain	TARIFA	G	0100	0500	0900	1300	1700	2100	(English and Spanish)
SE Spain	CABO LA NAO	Z	(TIMES TO BE DECIDED)						(English and Spanish)
S France	CROSS LA GARDE	(TBD)	(TIMES TO BE DECIDED)						
Sardinia	CAGLIARI	T	(PLANNED FOR 1994)						(English and Italian)
Sicily	AUGUSTA	S	(PLANNED FOR 1993)						(English and Italian)
Malta	MALTA	O	0220	0620	1020	1420	1820	2200	
W Italy	ROMA	R	(PLANNED FOR 1995)						(English and Italian)
NE Italy	ANCONA	U	(PLANNED FOR 1995)						(English and Italian)
Croatia	SPLIT	Q	0250	0650	1050	1450	1850	2250	
E Greece	KERKYRA	K	0140	0540	0940	1340	1740	2148	(English and Greek)
W Greece	LIMNOS	L	0150	0550	0950	1350	1750	2150	(English and Greek)
Crete	IRAKLION	H	0110	0510	0910	1310	1710	2110	(English and Greek)
W Turkey	IZMIR	I	0120	0520	0920	1320	1720	2120	
S Turkey	ANTALYA	F	0050	0450	0850	1250	1650	2050	
Cyprus	TROODOS	M	0200	0600	1000	1400	1800	2200	
Israel	HAIFA	P	0020	0420	0820	1220	1620	2020	(TO BE CONFIRMED)
Egypt	ISMAILIA	N	(TIMES NOT KNOWN)						
Egypt	ALEXANDRIA	N	(BEING RE-EQUIPPED - SERVICE MEANWHILE FROM ISMALIA)						

(BLACK SEA/BOSPHOROS/SEA OF MARMARA)

LOCATION	STATION	STATION IDENT	BROADCAST TIMES						REMARKS
Turkey	ISTANBUL	D	0030	0430	0830	1230	1630	2030	
Bulgaria	VARNA	J	0130	0530	1730				Weather Forecast
			0930	1330	2130				
Ukraine	ODESSA	C	0230	0630	1430	2230			
			1030	1830					Weather Forecast
			1830						(Includes Ice Report)
Ukraine	MARIUPOL	B	0100	1300	2100				
			0500	1700					Weather Forecast
			0900						Ice Report
Georgia	NOVOROSSIYSK	A	0300	0700	1500	2300			
			1100	1900					Weather Forecast
			1900						(Includes Ice Report)
Turkey	SAMSUN	E	0040	0440	0840	1240	1640	2040	

LOCATION	STATION	STATION IDENT	BROADCAST TIMES						REMARKS
NAVAREA IV									
NW Canada	CARTWRIGHT	(TBD)							
NW Canada	SEPT-ILES	(TBD)							
ST Lawrence	MONTREAL	(TBD)							
L Huron	WIARTON	(TBD)							
L Superior	THUNDERBAY	(TBD)							
Nova Scotia	SYDNEY	Q/0040	0540	0940	1340	1740	2140		
	YARMOUTH	(TBD)							
Massachusetts	BOSTON	F	0045	0445	0845	1245	1645	2045	
Massachusetts	BOSTON	K							
Virginia	PORTSMOUTH	N	0130	0530	0930	1330	1730	1930	
Bermuda	ST GEORGES	B	0100	0700	1300	1900			(Planned)
Florida	MIAMI	A	0000	0400	0800	1200	1600	2000	
Gulf Coast	NEW ORLEANS	G	0300	0700	1100	1500	1900	2100	
Puerto Rico	SAN JUAN	R	0200	0600	1000	1400	1800	2200	

NAVAREA V

(No stations active/planned area five. Brazil intends to operate International NAVTEX, including a 4MHz service)

NAVAREA VI

LOCATION	STATION	STATION IDENT	BROADCAST TIMES				REMARKS
Uruguay	SALTO	(TBD)					
Uruguay	LA PALOMA	(TBD)					
Uruguay	PUNTA DEL ESTE	(TBD)					
Uruguay	LAGUNA DEL SAUCE	(TBD)					
Uruguay	MONTIVIDEO	(TBD)					
Uruguay	COLONIA	(TBD)					
Argentina	ROSARIO	G	0110	0610	1210	1810	
Argentina	BUENOS AIRES	F	0510	1110	1710	2310	(Planned) (English and Spanish)
Argentina	MAR DEL PLATA	E	0110	0710	1310	1910	
Argentina	BAHIA BLANCA	D	0210	0810	1410	2010	(Planned)
Argentina	COMODORO RIVADAVIA	C	0040	0640	1240	1840	(Planned)
Argentina	RIO GALLEGOS	B	0140	0740	1340	1940	
Argentina	USHUAIA	A	0240	0840	1440	2040	

NAVAREA VII

(No stations active/planned area seven)

LOCATION	STATION	STATION IDENT	BROADCAST TIMES						REMARKS
NAVAREA VIII									
NW India	BOMBAY	G	0100	0500	1300	1700			(Planned for 4209.5kHz)
			0900	2100					Weather Forecast
E India	MADRAS	P	0230	0630	0930	1430	1830	2230	
NAVAREA IX									
Suez	ISMALIA	X	(TO BE DECIDED)						
Red Sea	JEDDAH	H	(TO BE DECIDED)						
Persian Gulf	DAMMAM	G	(TO BE DECIDED)						
Bahrain	HAMALA	B	0010	0410	0810	1210	1610	2010	

NAVAREA X

(No stations active/planned area ten)

LOCATION	STATION	STATION IDENT	BROADCAST TIMES						REMARKS
NAVAREA XI									
Guam	GUAM	V							
Indonesia	JAYAPURA	A	0000	0400	0800	1200	1600	2000	
Indonesia	AMBON	B	0010	0410	0810	1210	1610	2010	
Indonesia	MAKASSAR	D	0030	0430	0830	1230	1830	2030	
Indonesia	JAKARTA	E	0040	0440	0840	1240	1640	2040	
Singapore	JURONG	C	0020	0420	0820	1220	1420	2020	
Thailand	BANGKOK	(TBD)							
China	ZHANJIANG	M	(TO BE DECIDED)						English and Chinese
Hong Kong	HONG KONG	L	0150	0550	0950	1350	1750	2150	
China	GUANGZHOU	N	0210	0610	1010	1410	2210		English and Chinese
China	FUZHOU	O	(TO BE DECIDED)						English and Chinese
China	SHANGHAI	Q	0240	0640	1040	1440	2240		English and Chinese
China	TIANJIN	S	(TO BE DECIDED)						English and Chinese
China	DALIAN	R	0250	0650	1050	1450	2250		English and Chinese
Rep. of Korea	(TBD)	(TBD)							
Japan	OTARU	J	0130	0530	0930	1330	1730	2130	
Japan	KUSHIRO	K	0140	0540	0940	1340	1740	2140	
Japan	YOKOHAMA	I	0120	0520	0920	1320	1720	2120	
Japan	MOJI	H	0110	0510	0910	1310	1710	2120	
Japan	NAHA	G	0100	0500	0900	1300	1700	2100	

LOCATION	STATION	STATION IDENT	BROADCAST TIMES						REMARKS
NAVAREA XII									
Alaska	ADAK	X	0340	0740	1140	1540	1940	2340	
Alaska	KODIAK	J	0300	0700	1100	1500	1900	2300	
Canada	PRINCE RUPERT	(TBD)							
Canada	TOFINO	(TBD)							
Oregon	ASTORIA	W	0130	0530	0930	1330	1730	2130	
California	POINT REYES	C	0000	0400	0800	1200	1600	2000	
California	LONG BEACH	Q	0045	0445	0845	1245	1645	2045	
Hawaii	HONOLULU	O	0040	0440	0840	1240	1640	2040	
NAVAREA XIII									
Russia	VLADIVOSTOK	A	0000	0400	0800	1200	1600	2000	(Experimental 1990)
Russia	KHOLMSK	B	0010	0410	0810	1210	1610	2010	(as above)
Russia	PETROPAVLOVSK	C	0020	0420	0820	1220	1620	2020	(as above)
Russia	MAGADAN	D	0030	0430	0830	1230	1630	2030	(as above)
Russia	BERINGOVSKIY	E	0040	0440	0840	1240	1640	2040	(as above)
Russia	PROVIDENYA	F	0050	0450	0850	1250	1650	2050	(as above)
NAVAREA XIV									
(No stations active/planned for area fourteen)									
NAVAREA XV									
Chile	ANTOFAGASTA	A	0000	0400	0800	1200	1600	2000	English and Spanish
Chile	VALPARISO	B	0010	0410	0810	1210	1610	2010	English and Spanish
Chile	TALCAHUANO	C	0020	0420	0820	1220	1620	2020	English and Spanish
Chile	PUERTO MONTT	D	0030	0430	0830	1230	1630	2030	English and Spanish
Chile	PUNTA ARENAS	E	0040	0440	0840	1240	1640	2040	English and Spanish
NAVAREA XVI									
Peru	PAITA	S	0300	0700	1100	1500	1900	2300	(Under trial)
Peru	CALLAO	U	0320	0720	1120	1520	1920	2320	(Under trial)
Peru	MOLLENDO	W	0340	0740	1140	1540	1940	2340	(Under trial)

Appendix K

SHORT-WAVE BROADCAST (NEWS & ENTERTAINMENT) STATIONS

The following table lists a number of frequencies and stations which you can find within the shortwave (high frequency) broadcast bands. The stations listed form a sample only and, as you scan the broadcast bands, you will come across a large number of additional stations - broadcasting in a number of languages.

The stations listed all broadcast in the English language, and some also broadcast in other languages. Programmes vary from those directed at the international audience and those for nationals of the broadcasting nation, abroad.

Not all frequencies listed are in use throughout the day/night, but are chosen by the broadcast station according to the area of the world they are targeting at any particular time. Some stations also use directional antennas to achieve this, which will result in a much reduced signal outside of the directed region.

As the first edition of this manual goes print (Spring 1994) we are approaching a "low" in the (approximately) 11-year sunspot cycle. Shortwave reception over long distances/at higher frequencies will therefore be at its worst for some time. However successful you are in picking up distant stations - it can only get better as the sunspot activity hots up over the coming years (1997 onwards).

Many stations alter programme times and content twice yearly and, for this reason, I have not attempted to show programme details (which can be obtained direct from the broadcaster concerned - addresses in Appendix Q).

The sample stations listed are:
BBC (British Broadcasting Corporation World Service);
RCA (Radio Canada International)
RNZI (Radio New Zealand International)
ROZ (Radio Australia);
SRI (Swiss Radio International)
VOA (Voice of America)

Freq'y (kHz)	Station	Freq'y (kHz)	Station	Freq'y (kHz)	Station	Freq'y (kHz)	Station
621	VOA	1197	VOA	3255	BBC	5880	ROZ
648	BBC	1260	VOA	3915	BBC	5960	RCI
792	BBC	1296	BBC	3955	BBC	5965	BBC
792	VOA	1413	BBC	3980	VOA	5975	BBC
930	VOA	1530	VOA	3985	SRI	5985	VOA
1143	VOA	1575	VOA			5995	RCI
1197	BBC	1580	VOA			5995	ROZ

Freq'y (kHz)	Station	Freq'y (kHz)	Station	Freq'y (kHz)	Station	Freq'y (kHz)	Station
5995	VOA	7205	VOA	9650	SRI	11750	BBC
		7215	BBC	9660	BBC	11760	BBC
6005	BBC	7230	RCI	9670	BBC	11760	VOA
6005	VOA	7235	RCI	9670	RCI	11765	BBC
6020	ROZ	7240	ROZ	9675	RNZI	11800	ROZ
6030	SRI	7260	ROZ	9700	BBC	11805	VOA
6035	VOA	7265	VOA	9700	RNZI	11820	BBC
6040	VOA	7280	VOA	9700	VOA	11820	VOA
6050	RCI	7325	BBC	9710	ROZ	11825	VOA
6060	VOA	7405	VOA	9740	BBC	11835	VOA
6080	ROZ	7480	SRI	9740	RCI	11845	RCI
6090	VOA	7625	VOA	9740	VOA	11850	VOA
6110	VOA			9750	BBC	11855	RCI
6130	VOA	9160	RCI	9755	RCI	11855	ROZ
6135	SRI	9410	BBC	9760	BBC	11855	VOA
6140	VOA	9455	VOA	9760	RCI	11860	BBC
6150	RCI	9505	RCI	9760	ROZ	11870	VOA
6155	VOA	9510	RNZI	9760	VOA	11880	RCI
6160	VOA	9510	ROZ	9770	VOA	11880	ROZ
6165	SRI	9515	BBC	9775	VOA	11895	VOA
6175	BBC	9525	VOA	9810	SRI	11910	ROZ
6180	BBC	9530	VOA	9860	SRI	11915	RCI
6180	VOA	9535	RCI	9870	ROZ	11915	VOA
6190	BBC	9535	SRI	9870	ROZ	11920	VOA
6195	BBC	9555	RCI	9885	SRI	11925	VOA
		9535	RCI	9885	VOA	11935	RCI
7105	BBC	9540	ROZ	9915	BBC	11940	BBC
7115	VOA	9560	SRI			11940	RCI
7120	VOA	9570	BBC	11580	VOA	11945	BBC
7125	VOA	9575	VOA	11670	ROZ	11945	RCI
7135	BBC	9580	BBC	11690	SRI	11955	BBC
7140	VOA	9580	ROZ	11695	VOA	11955	RCI
7150	BBC	9590	BBC	11705	RCI	11960	VOA
7150	RCI	9590	VOA	11705	VOA	11995	VOA
7155	RCI	9600	BBC	11715	VOA		
7160	BBC	9610	BBC	11720	ROZ	12035	SRI
7170	VOA	9630	BBC	11720	VOA	12095	BBC
7180	BBC	9640	BBC	11730	BBC		
7180	RCI	9645	VOA	11730	RCI	13605	ROZ
7195	VOA	9650	RCI	11735	VOA	13635	SRI

Freq'y (kHz)	Station	Freq'y (kHz)	Station	Freq'y (kHz)	Station
13650	RCI	15380	BBC	21470	BBC
13670	RCI	15395	VOA	21485	VOA
13685	SRI	15400	BBC	21545	RCI
13710	VOA	15410	VOA	21550	VOA
13720	RCI	15420	BBC	21660	BBC
13730	ROZ	15425	VOA	21675	RCI
13755	ROZ	15430	SRI	21715	BBC
		15445	VOA	21725	ROZ
15070	BBC	15495	VOA	21740	ROZ
15105	BBC	15505	SRI	21770	SRI
15115	VOA	15575	BBC	21820	SRI
15120	RNZI	15580	VOA		
15120	VOA	15600	VOA	25750	ROZ
15140	RCI	17565	SRI		
15150	RCI	17635	SRI		
15155	VOA	17640	BBC		
15160	VOA	17695	ROZ		
15170	ROZ	17670	SRI		
15180	VOA	17705	BBC		
15185	VOA	17715	ROZ		
15195	VOA	17730	SRI		
15205	BBC	17735	VOA		
15205	VOA	17740	VOA		
15220	BBC	17750	ROZ		
15240	ROZ	17770	RNZI		
15245	VOA	17770	VOA		
15250	VOA	17790	BBC		
15255	VOA	17795	ROZ		
15260	BBC	17800	VOA		
15260	RCI	17820	RCI		
15280	BBC	17820	VOA		
15290	VOA	17830	BBC		
15305	VOA	17840	BBC		
15310	BBC	17860	BBC		
15320	ROZ	17875	RCI		
15320	VOA	17880	BBC		
15325	RCI	17880	ROZ		
15340	BBC	17885	BBC		
15360	BBC				
15365	ROZ	21455	VOA		

AMATEUR SERVICE BANDS (SHORT- WAVE/MEDIUM- WAVE)
'Worldband', 'Broadcast' and Marine receivers which cover the short-wave bands can be used to monitor Amateur Service bands. Marine receivers may be configured to use the ITU Channels when selecting (maritime) frequencies - if that is the case with your own receiver, you need to go to the 'continuous tune' (or equivalent) mode when listening across the amateur and/or broadcast bands.

Amateur bands cover much more than is shown below - only the short-wave/medium - wave bands which can be picked-up with a normal marine SSB set are covered here. (The intention, as with other Appendices, is to give you the reader an insight to what is available - not to provide comprehensive coverage, which would require another complete book).

Metric Annotation **'Meter Band'**	*Frequencies covered in each ITU Region (kHz)*			**Remarks**
	Region One	**Region Two**	**Region Three**	
160m (One-Sixty)	1810-1850	1800-2000	1800-2000	Different parts of this band are shared with other Services (eg, Radiolocation) in each of the three Regions. 'Phone starts at 1840kHz, all three Regions.
80m (Eighty)	3500-3800	3500-4000	3500-3900	Shared with Fixed, Aeronautical and Land Mobile, and Broadcast Services (especially towards the top end of the Band). 'Phone starts at 3525kHz (Region 2), 3535kHz (Region 3) and 3590kHz (Region 1).
40m (Forty)	7000-7100	7000-7300	7000-7100	7100-7300 is allocated to the Broadcast Service in Regions One and Three, which can make listening difficult for Amateurs in Region Two. Phone above 7030 (Region 3) and above 7040 (1 and 2).

Metric Annotation	*Frequencies covered in each ITU Region (kHz)*			
'Meter Band'	**Region One**	**Region Two**	**Region Three**	**Remarks**
30m (Thirty)	10100-10150	10100-10150	10100-10150	No 'Phone' Band in any Region. Fixed Service use in Regions One/Three limit the usefulness of this Band.
20m (Twenty)	14000-14350	14000-14350	14000-14350	This is the 'cleanest' Band, around the world - offering reasonable to good propagation throughout the sunspot cycle and with minimal interference from other Services. Phone band starts at 14101, all three Regions.
17m (Seventeen)	18068-18168	18068-18168	18068-18168	Mainly Morse and RTTY below 18110kHz. Phone/Morse above, in all three Regions.
15m (Fifteen)	21000-21450	21000-21450	21000-21450	Beacons around 21149-21151; Morse/RTTY below; 'Phone and Morse above. Good long range band when sunspot activity high.
12m (Twelve)	24890-24990	24890-24990	24890-24990	Phone above 24930kHz (Regions 1/2) and above 24920 (Region 3). Needs higher sunspot activity.
10m (Ten)	28000-29700	28000-29700	28000-29700	'Phone (with Morse, SSTV, repeaters, satellite) above 28300; otherwise Beacons, Morse, RTTY and packet.

The tendency within each Band is for Morse, Teletype etc to be towards the lower end of the band with 'Phone' towards the higher end. Access to the various parts of the Bands, within each Region, is also restricted according to the level of 'ticket' held, with Novices having access to the smallest section, and 'Extra' having access to the complete Band. 'RTTY' - originally an abbreviation for 'Radio-teletype' is now used to cover AMTOR (Telex Over Radio, like SITOR in the Marine Bands); and Baudot (ITA2 signalling), ASCII and Packet Radio.

AMATEUR RADIO SERVICE – MARITIME MOBILE NETS

Days	Time (UTC)	Freq (kHz)	Net Name	Area Covered
Dy	0000	7158	Caribbean Net	Caribbean Sea
Dy	0000	14313	International MM Net	World-wide
Dy	0130	28313	S Calif' 10m Maritime	World-wide
Dy	0300	28480	Ten Meter MM Net	Pacific Rim Nations and Waters
Dy	0400	14115	DDD Net-Pacific for Canadian	Pacific Ocean
Dy	0400	14310	Maritime Emergency Net	NE Canada
Dy	0400	14314	Pacific MM Net	Pacific Ocean
Dy	0530	14303	Swedish Maritime Net	Pacific Ocean
Dy	0700	14310	Mariana - Guam	Pacific Ocean
Dy	0715	3815	South Pacific Net	South Pacific
Dy	0800	14303	UK Net	Pacific Ocean
Dy	0800	14315	South Pacific Net	South Pacific
Dy	1100	3750	Maritime WX Net	NE Canada
Dy	1100	7230	Caribbean MM Net	Caribbean Sea
S,Su	1200	28380	Persian Gulf Rag Chew Net	World-wide
Dy	1245	7268	Waterway Radio and Cruising Club Net	US East Coast, Atlantic, Caribbean
Dy	1300	21400	Trans Atlantic MM Net	N/S Atlantic & Caribbean
Dy	1400	3968	SONRISA	W Coast Mexico & Sea of Cortez
Dy	1530	7294	Chubasco Net	W USA & W Coast Mexico
Dy	1600	7238.5	Baja Calif' Maritime Net	South Calif; W Mexico; Sea of Cortez
Dy	1630	14303	Swedish Maritime Net	Indian Ocean
M-F	1630	14340	California/Hawaii Net	W Coast/Pacific Ocean
M-F	1730	14115	DDD Net-Pacific for Canadians	Pacific Ocean
Dy	1800	14313	Maritime Mobile Service Net	
Dy	1800	14303	UK Net	Atlantic Ocean
M-F	1900	14305	Confusion Net	Pacific Ocean
M-S	1900	14340	Mariana Net	Pacific Ocean
Dy	2030	14303	Swedish Maritime Net	Atlantic Ocean
Dy	2230	7190	W Coast Admirals MM Net	US
M-F	2230	21404	Pacific MM Net	Pacific Ocean
M	2300	14285	California to Caribbean	Caribbean Sea
M	2310	14285	California to S Pacific	Pacific Ocean

1 Dy = Daily; S = Sat; Su = Sun; M-F = Mon -Fri, etc

2 MM = Maritime Mobile

3 WX = Weather

4 The frequencies listed are not assigned to the net, but have been chosen by practising MM Hams as being good for the purpose. Some may change day-by-day because of propagation conditions/interference, by a few kiloHertz, or even onto another band.

5 MM nets are only a small proportion of the total number of nets world-wide, and mariners are not barred from working non-MM nets. The 'ARRL Net Directory' lists a large number of nets, for different purposes.

The above table was reproduced with permission from the *ARRL Net Directory*; © ARRL.

World Time Zones

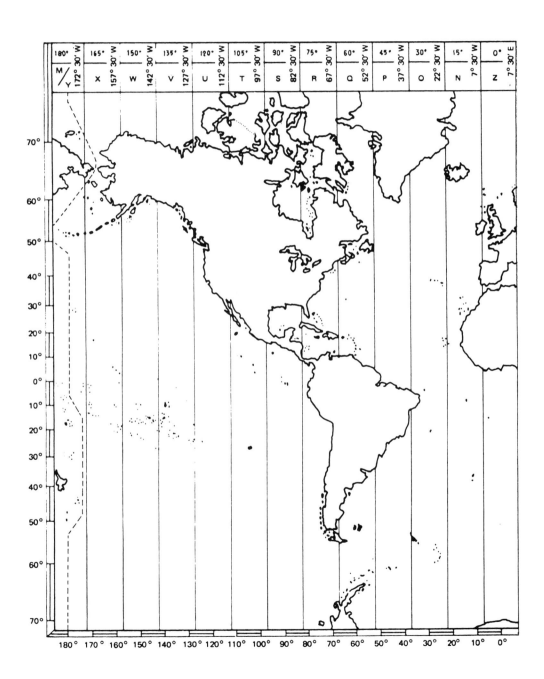

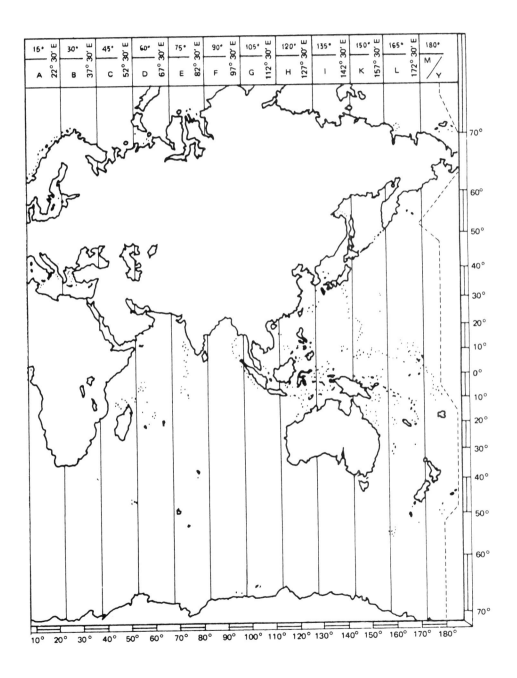

HOURS BEHIND UTC

-12 M/Y	-11 X	-10 W	-9 V	-8 U	-7 T	-6 S	-5 B	-4 Q	-3 P	-2 O	-1 N	0 Z
1200	1300	1400	1500	1600	1700	1800	1900	2000	2100	2200	2300	0000
13	14	15	16	17	18	19	20	21	22	23	00	0100
14	15	16	17	18	19	20	21	22	23	00	01	0200
15	16	17	18	19	20	21	22	23	00	01	02	0300
16	17	18	19	20	21	22	23	00	01	02	03	0400
17	18	19	20	21	22	23	00	01	02	03	04	0500
18	19	20	21	22	23	0C	01	02	03	04	05	0600
19	20	21	22	23	00	01	02	03	04	05	06	0700
20	21	22	23	00	01	02	03	04	05	06	07	0800
21	22	23	0C	01	02	03	04	05	06	07	08	0900
22	23	00	01	02	03	04	05	06	07	08	09	1000
23	00	01	02	03	04	05	06	07	08	09	10	1100
00	0100	0200	0300	0400	0500	0600	0700	0800	0900	1000	1100	1200
01	02	03	04	05	06	07	08	09	10	11	12	1300
02	03	04	05	06	07	08	09	10	11	12	13	1400
03	04	05	06	07	08	09	10	11	12	13	14	1500
04	05	06	07	08	09	10	11	12	13	14	15	1600
05	06	07	08	09	10	11	12	13	14	15	16	1700
06	07	08	09	10	11	12	13	14	15	16	17	1800
07	08	09	10	11	12	13	14	15	16	17	18	1900
08	09	10	11	12	13	14	15	16	17	18	19	2000
09	10	11	12	13	14	15	16	17	18	19	20	2100
10	11	12	13	14	15	16	17	18	19	20	21	2200
11	12	13	14	15	16	17	18	19	20	21	22	2300
12	13	14	15	16	17	18	19	20	21	22	23	2400

NOTES: UTC does not change and is the normal 'wintertime' of the United Kingdom.
The UK, like many other countries, operates 'summertime' from around the end of March

HOURS AHEAD OF UTC

0 Z	+1 A	+2 B	+3 C	+4 D	+5 E	+6 F	+7 G	+8 H	+9 I	+10 K	+11 L	+12 M/Y
0000	0100	0200	0300	0400	0500	0600	0700	0800	0900	1000	1100	1200
0100	02	03	04	05	06	07	08	09	10	11	12	13
0200	03	04	05	06	07	08	09	10	11	12	13	14
0300	04	05	06	07	08	09	10	11	12	13	14	15
0400	05	06	07	08	09	10	11	12	13	14	15	16
0500	06	07	08	09	10	11	12	13	14	15	16	17
0600	07	08	09	10	11	12	13	14	15	16	17	18
0700	08	09	10	11	12	13	14	15	16	17	18	19
0800	09	10	11	12	13	14	15	16	17	18	19	20
0900	10	11	12	13	14	15	16	17	18	19	20	21
1000	11	12	13	14	15	16	17	18	19	20	21	22
1100	12	13	14	15	16	17	18	19	20	21	22	23
1200	1300	1400	1500	1600	1700	1800	1900	2000	2100	2200	2300	2400
1300	14	15	16	17	18	19	20	21	22	23	24	01
1400	15	16	17	18	19	20	21	22	23	24	01	02
1500	16	17	18	19	20	21	22	23	24	01	02	03
1600	17	18	19	20	21	22	23	24	01	02	03	04
1700	18	19	20	21	22	23	24	01	02	03	04	05
1800	19	20	21	22	23	24	01	02	03	04	05	06
1900	20	21	22	23	24	01	02	03	04	05	06	07
2000	21	22	23	24	01	02	03	04	05	06	07	08
2100	22	23	24	01	02	03	04	05	06	07	08	09
2200	23	24	01	02	03	04	05	06	07	08	09	10
2300	24	01	02	03	04	05	06	07	08	09	10	11
2400	01	02	03	04	05	06	07	08	09	10	11	12

to the end of October. You need to refer to a current Almanac (or other source) for summertime details in your current area of operation.

Use the sheet below to plan your daily watchkeeping schedule. This is especially important where (i) you are working in local time and broadcasts are being publicised in UTC and (ii) you need to use the same radio equipment for more than one purpose (eg – your SSB set for weatherfax and an Inter-ship Schedule).

WATCHKEEPING SCHEDULE FOR.. (DATE)   AM

Period	Enter Time Local	UTC	Station(s) Required	Frequencies Available	Remarks (eg Wefax)	Equipment Needed	Sig Strength & Readability :
00-01							
01-02							
02-03							
03-04							
04-05							
05-06							
06-07							
07-08							
08-09							
09-10							
10-11							
11-12							

WATCHKEEPING SCHEDULE FOR.. (DATE)   PM

Period	Enter Time Local	UTC	Station(s) Required	Frequencies Available	Remarks (eg Wefax)	Equipment Needed	Sig Strength & Readability
12-13							
13-14							
14-15							
15-16							
16-17							
17-18							
18-19							
19-20							
20-21							
21-22							
22-23							
23-24							

WORLD MARI-TIME ZONES

Showing hours ahead (+) and behind (-) Co-ordinated Universal Time (UTC) for Standard and Daylight Saving Time (DST) for selected locations around the world.

Location	Standard	DST	Location	Standard	DST
Alaska	-9	-8	Manitoba	-6	-5
	-10	-9	Northern Territories	-7	-6
Albania	+1	+2	Yukon & BC	-8	-7
Algeria	+1		Cambodia	+7	
Angola	+1		Canary Islands	UTC	+1
Anguilla	-4		Cape Verde Is.	-1	
Antigua	-4		Cayman Is.	-5	-4
Argentina (East)	-3	-2	Chile	-4	-3
Argentina	-3		China		
Aruba	-4		Shanghai	+8	+9
Ascension Island	UTC		Christmas Is.	+7	
Australia:			Cocos Island	+6.5	
Queensland	+10		Colombia	-5	
Victoria, NSW	+10		Comora Rep.	+3	
Tasmania	+10	+11	Congo	+1	
S Australia (East)	+9.5	+10.5	Cook Is.	-10	-9.5
S Australia (West)	+9	+10	Costa Rica	-6	-5
W Australia	+8		Croatia	+1	+2
Azerbaidzhan	+4	+5	Cuba	-5	-4
Azores	-1	UTC	Cyprus	+2	+3
			Czech & Slovak Rep.	+1	+2
Bahamas	-5	-4			
Bahrain	+3		Denmark	+1	+2
Bangladesh	+6		Diego Garcia	+5	
Barbados	-4		Djibouti	+3	
Belarus	+2	+3	Dominica	-4	
Belgium	+1	+2	Dom. Rep.	-4	
Belize	-6				
Benin	-6		Easter Is.	-6	-5
Bermuda	-4	-3	Ecuador	-5	
Boznia/Hercegovina	+1	+2	Egypt	+2	
Brazil			El Salvador	-6	
Oceanic Islands	-2		Equatoria Guinea	+1	
East/Coastal	-3	-2	Estonia	+2	+3
Manaos	-4	-3	Ethiopia	+3	
Acre	-5	-4			
Brunei	+8	+8	Falkland Is.	-4	
Bulgaria	+2	+3	(Port Stanley)	-4	-3
			Faroe Is.	UTC	+1
Cameroon	+1		Fiji	+12	
Canada			Finland	+2	+3
SE Labrador &			France	+1	+2
Newfoundland	-3.5	-2.5			
Eastern Canada	-4	-3	Gabon	+1	
Ontario, Quebec	-5	-4	Gambia	UTC	

Location	Standard	DST	Location	Standard	DST
Georgia	+4	+5	Libya	+1	+2
Germany	+1	+2	Lithuania	+2	+3
Ghana	UTC		Lord Howe ls.	+10.5	+11
Gibraltar	+1	+2	Luxembourg	+1	+2
Greece	+2	+3			
Greenland			Macau	+8	
Scoresbysund	-1	UTC	Madagascar	+3	
Thule area	-3		Madeira	UTC	+1
Other areas	-3	-2	Malawi	+2	
Grenada	-4		Malaysia	+8	
Guadeloupe	-4		Maldive ls.	+5	
Guam	+10		Malta	+1	+2
Guatemala	-6	-5	Marshall ls.	+12	
Guiana (French)	-3		Martinique	-4	
Guinea (Rep.)	UTC		Mauritania	UTC	
Guinea Bissau	UTC		Mauritius	+4	
Guyana (Rep.)	-3		Mayotte	+3	
			Mexico	-6	
Haiti	-5	-4			
Hawaii	-10		Micronesia		
Honduras (Rep.)	-6		Truk, Yap	+10	
Hong Kong	+8		Pohnpei	+11	
Hungary	+1	+2	Moldavia	+2	+3
			Monaco	+1	+2
Iceland	UTC		Montserrat	-4	
India	+5.5		Morocco	UTC	
Indonesia			Mozambique	+2	
Java, Bali,Sumatra	+7		Myanmar	+6.5	
Kalimantan, Sulawesi,					
Timor	+8		Nauru	+12	
Moluccas, We.lrian	+9		Netherlands	+1	+2
Iran	+3.5		Neth. Antilles	-4	
Iraq	+3	+4	New Caledonia	+11	
Ireland	UTC	+1	New Zealand	+12	+13
Israel	+2	+3	Nicaragua	-6	
Italy	+1	+2	Niger	+1	
Ivory Coast	UTC		Nigeria	+1	
			Niue	-11	
Jamaica	-5	-4	Norfolk ls.	+11.5	
Japan	+9		N. Marianas	+10	
Jordon	+2	+3	Norway	+1	+2
Kenya	+3		Oman	+4	
Korea (Rep.)	+9	+10			
Korea (DPR)	+9		Pakistan	+5	
Kuwait	+3		Palau	+9	
			Panama	-5	
Latvia	+2	+3	Papua N.Guinea	+10	
Lebanon	+2	+3	Peru	-5	-4
Lesotho	+2		Philippines	+8	
Liberia	UTC				

Location	Standard	DST	Location	Standard	DST
Poland	+1	+2	Tunisia	+1	+2
Polynesia (Fr.)	-10		Turks & Caicos	-4	
Portugal	UTC	+1	Turkey	+2	+3
Puerto Rico	-4		Tuvalu	+12	
Qatar	+3		Uganda	+3	
			Ukraine	+2	+3
Reunion	+4		United Arab Em.	+4	
Romania	+2	+3	United Kingdom	UTC	+1
Russia			Uruguay	-3	-2
Khabarovsk	+10	+11	USA		
Petropavlovsk	+12	+13	Eastern	-5	-4
Rwanda	+2		Indiana	-5	
			Central	-6	-5
Samoa Is.	-11		Mountain	-7	-6
S Tome	UTC		Arizona	-7	
Saudi Arabia	+3		Pacific	-8	-7
Senegal	UTC				
Seychelles	+4		Vanuatu	+11	+12
Sierra Leone	UTC		Venezuela	-4	
Singapore	+8		Vietnam	+7	
Slovenia	+1	+2	Virgin Is.	-4	
Solomon Is.	+11		Wake Is.	+12	
Somalia	+2		Wallis & Futuna	+12	
S. Africa	+2				
Spain	+1	+2	Yemen	+3	
			Former Yugoslavia	+1	+2
Sri Lanka	+5.5				
St Helena	UTC		Zaire (Kinshasa)	+1	
St Kitts-Nevis	-4		Zambia	+2	
St. Lucia	-4		Zimbabwe	+2	
St. Pierre	-3	-2			
St. Vincent	-4				
Sudan	+2				
Surinam	-3				
Swaziland	+2				
Sweden	+1	+2			
Switzerland	+1	+2			
Syria	+2	+3			
Taiwan	+8				
Tanzania	+3				
Thailand	+7				
Togo	UTC				
Tonga	+13				
Trinidad	-4				
Tristan da Cunha	UTC				

INMARSAT SATELLITE ALIGNMENT TABLES

Tables 1– 4 list Antenna Positioning data for the Inmarsat system. Each table shows an example of elevation and azimuth calculation to point your Satcom terminal to the chosen satellite, depending on your position relevant to that satellite. The general process is as shown below:

(1) Establish your vessel's latitude and longitude, to the nearest degree, using the available navigation equipment.

(2) Chose which satellite you wish to work through, relevant to your position in any one Ocean Region, (where you have a choice of satellite, you may want to choose the nearest — or you may choose one which works through your preferred CES).

(3) Estimate your vessel's position **relevant to the chosen satellite**, bearing in mind that for all four satellites the latitude is 0°; and that the longitude for each is:
AORW = 54° W; AOR-E = 15.5° W; IOR - 64.5° E; POR - 178° E.
You need to decide whether you are:
North and West; North and East; South and West; or South and East of the chosen bird. Depending on that relevant position, you will calculate the required Azimuth and Elevation from Tables P1 (N+W); P2 (N+E); P3 (S+W); or P4 (S+E).

(4) Calculate the difference between your vessel's longitude and the chosen satellite, to the nearest 5°. Don't forget to allow for East and West longitudes for satellite/vessel, where these are different — eg, if your vessel is in the Med at 10°E and you want to communicate through the Atlantic Ocean Region-East satellite at 15.5W, your difference in longitude is 10+15.5=25° (to the nearest 5°).

(5) Round the ships **latitude** the nearest 5°.

(6) Using the relevant table (P1-P4 following), trace your latitude on the vertical column and match with the difference in longitude on the horizontal scale. Read off the Azimuth (top scale) and the elevation (bottom scale). These are the initial settings required for your mobile antenna.

(7) Follow the manufacturers instructions for setting your antenna to the appropriate alignment.

NOTE:
(1) Some ship earth stations can set themselves — you need only key-in your correct position and the software calculates the settings and drives your antenna to the required alignment.

(2) Once set, a maritime antenna will allow for pitch, roll and position changes when your vessel is on the move — you only need to consider resetting after power-down.

(3) The tables show elevation settings down to zero degrees. In practice, elevations below 5 degrees (a/B) and 10 degrees (M) may prove unworkable. It should not be assumed that service is available at the extreme latitudes represented with the lower elevations included in the table. Similarly, for extreme longitudes for any Ocean Region, you would be better to work to a nearer satellite.

Appendix P1

ANTENNA ALIGNMENT - VESSEL LOCATED NORTH & WEST OF SATELLITE

Each cell shows Azimuth (upper figure) / Elevation (lower figure).

Ship's Lat °N	80	75	70	65	60	55	50	45	40	35	30	25	20	15	10	05	0
80	-	-	-	-	-	-	-	-	-	-	150/00	155/00	160/01	165/01	170/01	175/01	180/01
75	-	-	-	-	-	124/00	129/01	134/02	139/03	144/04	149/04	154/05	159/05	164/06	169/06	175/06	180/06
70	-	-	-	114/00	118/01	123/03	128/04	133/05	138/07	143/07	148/09	154/09	159/10	164/11	169/11	175/11	180/11
65	-	-	-	113/02	118/04	122/05	127/07	132/09	137/10	142/12	148/13	153/14	158/15	164/16	169/16	174/17	180/17
60	-	-	107/01	112/04	117/06	121/08	126/10	131/12	136/14	141/16	146/17	152/19	157/20	163/21	168/21	174/22	180/22
55	-	102/00	107/03	111/05	115/08	120/11	125/13	129/16	134/18	139/20	145/22	150/23	156/25	162/26	168/27	174/27	180/27
50	-	102/01	106/04	110/07	114/10	118/13	123/16	127/19	132/21	138/24	143/26	149/28	155/30	161/31	167/32	173/32	180/33
45	-	101/02	104/05	108/09	112/12	116/16	121/19	125/22	130/25	135/28	140/30	147/33	153/34	159/36	166/37	173/38	180/38
40	-	100/03	103/07	107/10	110/14	114/18	118/21	123/25	127/28	133/31	138/34	144/37	150/38	157/41	165/43	172/43	180/44
35	-	099/04	102/08	105/12	108/16	112/20	116/24	120/28	124/31	129/35	135/38	141/41	148/44	155/46	163/48	171/49	180/49
30	095/00	098/04	100/09	103/13	106/17	109/22	113/26	117/30	121/34	126/38	131/42	137/46	144/49	152/51	161/53	170/54	180/55
25	094/00	096/05	099/09	101/14	104/19	106/23	110/28	113/33	117/37	121/41	126/46	132/50	139/53	148/56	157/59	168/60	180/61
20	093/01	095/05	097/10	099/15	101/20	103/25	106/30	109/34	112/39	116/44	121/49	126/53	133/57	142/61	153/64	166/66	180/66
15	093/01	094/06	095/11	097/16	098/21	100/26	102/31	105/36	107/41	110/46	114/51	119/56	125/61	134/65	146/69	161/71	180/72
10	092/01	093/06	094/11	095/16	096/21	097/27	098/32	100/37	102/43	104/48	107/53	110/59	116/64	123/69	135/73	153/77	180/78
5	091/01	091/06	092/11	092/17	093/22	093/27	094/32	095/38	096/43	097/49	099/55	101/60	103/66	108/71	116/77	135/82	180/85
0	090/02	090/07	090/12	090/17	090/22	090/28	090/33	090/38	090/44	090/49	090/55	090/61	090/66	090/72	090/78	090/84	180/90

Difference in degrees longitude between ship and satellite.

Azimuth (= upper figure)
Elevation (= lower figure)

Ship's
Lat.
Degs.
North

ANTENNA ALIGNMENT - VESSEL LOCATED NORTH & EAST OF SATELLITE

Ship's Lat.		0	5	10	15	20	25	30	35	40	45	50	55	60	65	70	75	80
80	Az	180	185	190	195	200	205	210	-	-	-	-	-	-	-	-	-	-
	El	01	01	01	01	01	00	00	-	-	-	-	-	-	-	-	-	-
75	Az	180	185	190	196	201	206	211	216	221	226	231	236	-	-	-	-	-
	El	06	06	06	06	06	05	04	04	03	02	01	00	-	-	-	-	-
70	Az	180	185	191	196	201	206	212	217	222	227	232	237	242	246	-	-	-
	El	11	11	11	11	10	09	09	07	07	05	04	03	01	00	-	-	-
65	Az	180	186	191	196	202	207	212	128	223	228	233	238	242	247	-	-	-
	El	17	17	16	16	15	14	13	12	10	19	07	05	04	02	-	-	-
60	Az	180	186	192	197	203	208	214	219	224	229	234	239	243	248	253	-	-
	El	22	22	21	21	20	19	17	16	14	12	10	08	06	04	01	-	-
55	Az	180	186	192	198	204	210	215	221	226	231	235	240	245	249	253	258	-
	El	27	27	27	26	25	23	22	20	18	16	13	11	08	05	03	00	-
50	Az	180	187	193	199	205	211	217	222	228	233	237	242	246	250	254	258	-
	El	33	32	32	31	30	28	26	24	21	19	16	13	10	07	04	01	-
45	Az	180	187	194	201	207	213	220	225	230	235	239	244	248	252	256	259	-
	El	38	38	37	36	34	33	30	28	25	22	19	16	12	09	05	02	-
40	Az	180	188	195	203	210	216	222	227	233	237	242	246	250	253	257	260	-
	El	44	43	43	41	39	37	34	31	28	25	21	18	14	10	07	03	-
35	Az	180	189	197	205	212	219	225	231	236	240	244	248	252	255	258	261	-
	El	49	49	48	46	44	41	38	35	31	28	24	20	16	12	08	04	-
30	Az	180	190	199	208	216	223	229	234	239	243	247	251	254	257	260	262	265
	El	55	54	53	51	49	46	42	38	34	30	26	22	17	13	09	04	00
25	Az	180	192	203	212	221	226	234	239	243	247	250	254	256	259	261	264	266
	El	61	60	59	56	53	50	46	41	37	33	28	23	19	14	09	05	00
20	Az	180	194	207	218	227	234	239	244	248	251	254	257	259	262	263	265	267
	El	66	66	64	61	57	53	49	44	39	34	30	25	20	15	10	05	01
15	Az	180	199	214	226	235	241	246	250	253	255	258	260	262	263	265	266	267
	El	72	71	69	65	61	56	51	46	41	36	31	26	21	16	11	06	01
10	Az	180	207	225	237	244	250	253	256	258	260	262	263	264	265	266	267	268
	El	78	77	73	69	64	59	53	48	43	37	32	27	21	16	11	06	01
5	Az	180	225	244	252	257	259	261	263	264	265	266	267	267	268	268	269	269
	El	85	82	77	71	66	60	55	49	43	38	32	27	22	17	11	06	01
0	Az	180	270	270	270	270	270	270	270	270	270	270	270	270	270	270	270	270
	El	90	84	78	72	66	61	55	49	44	38	33	28	22	17	12	07	02

Ship's Lat. Degs. North

Difference in degrees longitude between ship and satellite

Azimuth (= upper figure)

Elevation (= lower figure)

ANTENNA ALIGNMENT - VESSEL LOCATED SOUTH & WEST OF SATELLITE

Each cell shows Azimuth (upper figure) / Elevation (lower figure).

Ship's Lat. Degs. South	80	75	70	65	60	55	50	45	40	35	30	25	20	15	10	5	0
0	090/02	090/07	090/12	090/17	090/22	090/28	090/33	090/38	090/44	090/49	090/55	090/61	090/66	090/72	090/78	090/84	000/90
5	089/01	089/06	088/11	088/17	087/22	087/27	086/32	085/38	084/43	083/49	081/55	079/60	077/66	072/71	064/77	045/82	000/85
10	088/01	087/06	086/11	085/16	084/21	083/27	082/32	080/37	078/43	076/48	073/53	070/59	064/64	057/69	045/73	027/77	000/78
15	087/01	086/06	085/11	083/16	082/21	080/26	078/31	075/36	073/41	070/46	066/51	061/56	055/61	046/65	034/69	019/71	000/72
20	087/01	085/05	083/10	081/15	079/20	077/25	074/30	071/34	068/39	064/44	059/49	054/53	047/57	038/61	027/64	014/66	000/66
25	086/00	084/05	081/09	079/14	076/19	074/23	070/28	067/33	063/37	059/41	054/46	048/50	041/53	032/56	023/59	012/60	000/61
30	085/00	082/04	080/09	077/13	074/17	071/22	067/26	063/30	059/34	054/38	049/42	043/46	036/49	028/51	019/53	010/55	000/55
35	-/-	081/04	078/08	075/12	072/16	068/20	064/24	060/28	056/31	051/35	045/38	039/41	032/44	025/46	017/48	009/49	000/49
40	-/-	080/03	077/07	073/10	070/14	066/18	062/21	057/25	053/28	047/31	042/34	036/37	030/39	023/41	015/43	008/43	000/44
45	-/-	079/02	076/05	072/09	068/12	064/16	059/19	055/22	050/25	045/28	039/30	033/33	027/34	021/36	014/37	007/38	000/38
50	-/-	078/01	074/04	070/07	066/10	062/13	057/16	053/19	048/21	042/24	037/26	031/28	025/30	019/31	013/32	007/32	000/33
55	-/-	078/00	073/03	069/05	065/08	060/11	055/13	051/16	045/18	041/20	035/22	030/23	024/25	018/26	012/27	006/27	000/27
60	-/-	-/-	072/01	068/04	063/06	059/08	054/10	049/12	044/14	039/16	034/17	028/19	023/20	017/21	012/21	006/22	000/22
65	-/-	-/-	-/-	067/02	062/04	058/05	053/07	048/09	043/10	038/12	032/13	027/14	022/15	016/16	011/16	006/17	000/17
70	-/-	-/-	-/-	066/00	062/01	057/03	052/04	047/05	042/07	037/07	032/09	026/09	021/10	016/11	011/11	005/11	000/11
75	-/-	-/-	-/-	-/-	-/-	056/00	051/01	046/02	041/03	036/04	031/04	026/05	021/05	016/06	010/06	005/06	000/06
80	-/-	-/-	-/-	-/-	-/-	-/-	-/-	-/-	-/-	-/-	030/00	025/00	020/01	015/01	010/01	005/01	000/01

Difference in degrees longitude between ship and satellite
Azimuth (= upper figure)
Elevation (= lower figure)

ANTENNA ALIGNMENT - VESSEL LOCATED SOUTH & EAST OF SATELLITE

Ship's Lat.	0	5	10	15	20	25	30	35	40	45	50	55	60	65	70	75	80
0 (Az)	360	270	270	270	270	270	270	270	270	270	270	270	270	270	270	270	270
(El)	90	84	78	72	66	61	55	49	44	38	33	28	22	17	12	07	02
5 (Az)	360	315	296	288	283	281	279	277	276	275	274	273	273	272	272	271	271
(El)	84	82	77	71	66	60	55	49	43	38	32	27	22	17	11	06	01
10 (Az)	360	333	315	303	296	290	287	284	282	280	278	277	276	275	274	273	272
(El)	78	77	73	69	64	59	53	48	43	37	32	27	21	16	11	06	01
15 (Az)	360	341	326	314	305	299	294	290	287	285	282	280	278	277	275	274	273
(El)	72	71	69	65	61	56	51	46	41	36	31	26	21	16	11	06	01
20 (Az)	360	346	333	322	313	306	301	296	292	289	286	283	281	279	277	275	273
(El)	66	66	64	61	57	53	49	44	39	34	30	25	20	15	10	05	01
25 (Az)	360	348	337	328	319	312	306	301	297	293	290	286	284	281	279	276	274
(El)	61	60	59	56	53	50	46	41	37	33	28	23	19	14	09	05	00
30 (Az)	360	350	341	332	324	317	311	306	301	297	293	289	286	283	280	278	275
(El)	55	55	53	51	49	46	42	38	34	30	26	22	17	13	09	04	00
35 (Az)	360	351	343	335	328	321	315	309	304	300	296	292	288	285	282	279	-
(El)	49	49	48	46	44	41	38	35	31	28	24	20	16	12	08	04	-
40 (Az)	360	352	345	337	330	324	318	313	307	303	298	294	290	287	283	280	-
(El)	44	43	43	41	39	37	34	31	28	25	21	18	14	10	07	03	-
45 (Az)	360	353	346	339	333	327	320	315	310	305	301	296	292	288	284	281	-
(El)	38	38	37	36	34	33	30	28	25	22	19	16	12	09	05	02	-
50 (Az)	360	353	347	341	335	329	323	318	312	307	303	298	294	290	286	282	-
(El)	33	32	32	31	30	28	26	24	21	19	16	13	10	07	04	01	-
55 (Az)	360	354	348	342	336	330	325	319	314	309	305	300	295	291	287	282	-
(El)	27	27	27	26	25	23	22	20	18	16	13	11	08	05	03	00	-
60 (Az)	360	354	348	343	337	332	326	321	316	311	306	301	297	292	287	-	-
(El)	22	22	21	21	20	19	17	16	14	12	10	08	06	04	01	-	-
65 (Az)	360	354	349	344	338	333	328	322	317	312	307	302	298	293	-	-	-
(El)	17	17	16	16	15	14	13	12	10	09	07	05	04	02	-	-	-
70 (Az)	360	355	349	344	339	334	328	323	318	313	308	303	298	294	-	-	-
(El)	11	11	11	11	10	09	09	07	07	05	04	03	01	00	-	-	-
75 (Az)	360	355	350	344	339	334	329	324	319	314	309	304	-	-	-	-	-
(El)	06	06	06	06	05	05	04	04	03	02	01	00	-	-	-	-	-
80 (Az)	360	355	350	345	340	330	210	-	-	-	-	-	-	-	-	-	-
(El)	01	01	01	01	01	00	00	-	-	-	-	-	-	-	-	-	-

Ship's Lat. Degs. South

Difference in degrees longitude between ship and satellite
Azimuth (= upper figure)
Elevation (= lower figure)

USEFUL ADDRESSES

The following list of addresses can be used to gain further information on such things as licensing (ship stations; maritime operator's permits; amateur stations; and amateur operators permits) — and information about specific types of service (maritime mobile; mobile satellite; amateur services; broadcast stations, etc).

AUSTRALIA - LICENSING
& OPERATORS CERTIFICATES:

Dept. of Transport & Communications
GPO Box 594,
CANBERRA
ACT 2601
Australia

CANADA - LICENSING
& OPERATORS CERTIFICATES:

Department of Communications
Radio Regulatory Branch,
300 Slater Street
OTTAWA, Ontario, K1A 0C8
CANADA

IRELAND - LICENSING
& OPERATORS CERTIFICATES:

Department of the Marine
Leeson Lane
DUBLIN 2
Ireland

NEW ZEALAND - LICENSING
& OPERATORS CERTIFICATES:

(or from RFS Field
Offices)

Communications Division
Radio Frequency Service
Ministry of Commerce Bldg
PO Box 2847
33 Bowen Street
WELLINGTON, New Zealand

SOUTH AFRICA - LICENSING
& OPERATORS CERTIFICATES:

Department of Posts & Telecommunications
Private Bag X74
PRETORIA 0001
Republic of South Africa

UK SHIP STATION LICENCE:

Radiocommunication Agency
Ship Radio Licensing Section
Room 613, Waterloo Bridge House,
Waterloo Road
LONDON SE1 8UA

UK GMDSS OPERATOR'S
CERTIFICATES; and
RESTRICTED RT CERT
(for SSB operation)

Wray Castle
Ambleside
Cumbria
LA22 0JB

UK RESTRICTED (VHF ONLY)
OPERATORS CERTIFICATE

ROYAL YACHTING ASSOCIATION
RYA House
Romsey Road
EASTLEIGH
HANTS SO5 4YA

USA SHIP STATION LICENCE:
& OPERATORS PERMITS;

FCC Private Radio Bureau
Licensing Division
Consumer Assistance Branch
1270 Fairfield Rd
Gettysburg
Pennsylvania 17325-7245

(Or from any FCC Field Office)

MOBILE SATELLITE COMMUNICATION SYSTEMS

INMARSAT Services:
(and service providers)

INMARSAT Maritime Division
INMARSAT HQ
99 City Road
LONDON EC1Y 1AX

BT-INMARSAT Services:
(for services through
Goonhilly CES)

BT INMARSAT CUSTOMER SERVICE
2ND Floor
43 Bartholomew Close
LONDON EC1A 7HP
Tel: (UK) 0800 378389
(Int) 44 71 492 4996

IRIDIUM MSS

IRIDIUM INC
(Director of Corporate Communications)
1350 L ST. NW, #500
WASHINGTON, DC 20005
USA

ODYSSEY MSS

TRW Space & Technology Group
Odyssey Program Office
One Space Park
Redondo Beach, CA 90278, USA

COMSAT

COMSAT Mobile Communications
Customer Support Dept.
22300 COMSAT Drive
Clarksburg, Maryland
20871, USA

Singapore Telecom Maritime Services
International Mobile Division
15 Hill Street 02-00
Telephone House
Singapore 0817

Telstra Maritime Services
(Radio/Satcom Services)
3/39 Herbert Street
St Leonards
NSW 2065, Australia

LONG-RANGE RADIOTELEPHONE SERVICES
(For Telstra Maritime see above)

AT&T High Seas
Radiotelephone Service
PO Box 8067
Trenton, NJ 08650-0067
USA

BT-Portishead Radio
Customer Service Dept
Highbridge, Somerset
TA9 3JY
England

Canadian Coastguard
Telecommunications &
Electronics Directorate
344 Slater Street
Canada, K1A ON7

Mobile Marine Radio Inc
Customer Service Dept
7700 Rinla Avenue
Mobile, AL 36619-1199
USA

AUTOMATIC RADIOTELEPHONE SERVICES

AUTOLINK RT
(Service details
- world-wide)

Global Maritime Radiotelephone Service
38 Bedford Place
Bloomsbury Square
LONDON WC1B 5JH

AUTOLINK RT
(Service details
- UK)

BT Maritime Radio Operations
3rd Floor
43 Bartholomew Close
LONDON EC1A 7HP

AUTOLINK RT
(Original Manufacturer)

Cimat SpA
Via Casalina Km. 22,900
00040 Montecompatri
Roma
ITALY

WATERCOM AMTS

Waterway Communications System Inc
453 East Park Place
JEFFERSONVILLE
Indiana 47130 USA
Tel: (800) 752 3000

Auto Seaphones
(Australia)
Contact Telstra Maritime

Telecom Mobile Radio
(Sealink Services)
78 Carbine Road
Mt Wellington
PO Box 6347
Wellesley Street
Auckland
New Zealand

AMATEUR RADIO SOCIETIES

AMERICAN RADIO RELAY LEAGUE, INC
225 Main Street
Newington,
Connecticut 06111
USA
Tel: +1 203 666 1541

WIRELESS INSTITUTE OF AUSTRALIA
National Society of Radio Amateurs
PO Box 300, South Caulfield,
Victoria. 3162

Tel: + 03 528 5962

RADIO SOCIETY OF GREAT BRITAIN
Lambda House
Cranborne Road
Potters Bar
Herts EN6 3JE
England/UK

NEW ZEALAND ASSOCIATION OF RADIO TRANSMITTERS
Inc
PO Box 40 525
Upper Hutt
NEW ZEALAND

IRISH RADIO TRANSMITTERS SOCIETY
PO Box 462
DUBLIN 9
IRELAND

SOUTH AFRICAN RADIO LEAGUE
PO Box 807
Houghton 2041
Republic of South Africa

THE CANADIAN RADIO RELAY LEAGUE, INC
RR #1
JORDAN, Ontario
L0R 1S0
CANADA

Tel: +1 519 660 1200

CANADIAN AMATEUR RADIO FEDERATION
PO Box 356
KINGSTON, Ontario
K7L 4W2
CANADA

Tel: +1 613 545 9100

RADIO AMATEUR DU QUEBEC INC. (RAQI)
4545 Pierre-de-Coubertin Avenue
PO Box 1000, Station 'M'
MONTREAL, Quebec
H1V 3R2

Tel: +1 514 252 3012

INTERNATIONAL BROADCAST STATIONS

World Radio TV Handbook
(Published annually, the WRTH can
be obtained through most good
bookshops. Contains details
of broadcast stations world-wide,
and useful items such as a world
propagation guide for the year)

BBC World Service
Bush House
PO Box 76
Strand
LONDON WC2B 4PH

Radio Australia (Overseas Service)
PO Box 755
Glen Waverly, Victoria 3150
Australia

Radio Canada International
PO Box 6000
Montreal
CANADA H3C 3A8

Radio France International
Societe Nationale de Radiodiffusion
116 Avenue du President-Kennedy
BP 9516
75016 PARIS

Radio New Zealand International
PO Box 2092
Wellington
New Zealand

SRI (Swiss Radio International)
P.O. Box, CH-3000 Berne 15,
Switzerland

Voice Of America
Office of Affiliate Relations
330 Independence Avenue, S.W.
Washington, D.C. 20547   USA

Index

active receive antenna, 54
Admiralty List of Radio Signals (ALRS), 69
amateur radio, 25,41,100,104
ambient voltage, 13
amplification, 15
antenna, 15,16,28,52,57,90,109,117
  alignment, 127
  cable, 59
  connector, 59
  coupling, 15
  gain, 57
  ground, 93
  Matching Unit (AMU/ATU), 18,82
  tuning unit, 15
  , active receive, 54
  , MF/HF, 18
  , quarter-wave, 18
  , receive, 19
  , VHF, 18
  , whip, 95
audio amplifier, 13,19,23
  frequency, 14
Autolink RT, 76,101
Automated Maritime Telephone Service, 76
automatic gain control, 23
  radiotelephone, 101

balanced modulator, 15

call-sign, 43,106
capture effect, 71
Carriage Requirements, 158
carrier frequency, 13,14
  generator, 13,15
Cellnet, 78
cellular radio, 78,141
Channel 16, 63,65,68,71,154
Citizens Band Radio, 74
class of emission, 39
co-axial cable, 59
Coast Earth Station (CES), 113

Coast Radio Station, 41
communications receiver, 100
compass, 62
Cospas-Sarsat, 73,123,128,149,160
counterpoise, 93

D-layer, 35,88
data terminal, 121
Decca, 31
detector, 23
Digital Selective Calling (DSC), 64,71,72,149
dipole, 57
Distress, 43,44,84,119,151
  frequency, 48
  traffic, 49
  transmission, 45
diversity reception, 145
Doppler effect, 130
dual-watch, 63
duplex, 41
DX contact, 105

E-layer, 35
earth mat, 82
electrolysis, 94
electromagnetic energy, 11,26
Emergency Position Indicating Radio Beacon (EPIRB), 50,72,123,136,154
Enhanced Group Call, 123
EPIRB registration, 134

F-layer, 35
FleetNET, 122
foghorn, 65
frequency modulation, 71
  multiplication, 13
  standard, 90

gain, 57
gang tune, 21